Assessing Revolutionary and Insurgent Strategies

UNDERSTANDING STATES OF RESISTANCE

Paul J. Tompkins Jr., USASOC Project Lead

W. Sam Lauber, Lead Author
Steven Babin, Katharine Burnett, Jonathon Cosgrove, Catherine
Kane, W. Sam Lauber, and Theodore Plettner, Contributing Authors

United States Army Special Operations Command
and
The Johns Hopkins University Applied Physics Laboratory

This publication is a work of the United States Government in accordance with Title 17, United States Code, sections 101 and 105.

Published by:

The United States Army Special Operations Command

Fort Bragg, North Carolina

Reproduction in whole or in part is permitted for any purpose of the United States government. Nonmateriel research on special warfare is performed in support of the requirements stated by the United States Army Special Operations Command, Department of the Army. This research is accomplished at the Johns Hopkins University Applied Physics Laboratory by the National Security Analysis Department, a nongovernmental agency operating under the supervision of the USASOC Sensitive Activities Division, Department of the Army.

The analysis and the opinions expressed within this document are solely those of the authors and do not necessarily reflect the positions of the US Army or the Johns Hopkins University Applied Physics Laboratory.

Comments correcting errors of fact and opinion, filling or indicating gaps of information, and suggesting other changes that may be appropriate should be addressed to:

United States Army Special Operations Command

G-3X, Sensitive Activities Division

2929 Desert Storm Drive

Fort Bragg, NC 28310

All ARIS products are available from USASOC at www.soc.mil under the ARIS link.

Published by Conflict Research Group.

First published by USASOC in 2019

CONFLICT
RESEARCH
GROUP

ASSESSING REVOLUTIONARY AND INSURGENT STRATEGIES

The Assessing Revolutionary and Insurgent Strategies (ARIS) series consists of a set of case studies and research conducted for the US Army Special Operations Command by the National Security Analysis Department of the Johns Hopkins University Applied Physics Laboratory.

The purpose of the ARIS series is to produce a collection of academically rigorous yet operationally relevant research materials to develop and illustrate a common understanding of insurgency and revolution. This research, intended to form a bedrock body of knowledge for members of the Special Forces, will allow users to distill vast amounts of material from a wide array of campaigns and extract relevant lessons, thereby enabling the development of future doctrine, professional education, and training.

From its inception, ARIS has been focused on exploring historical and current revolutions and insurgencies for the purpose of identifying emerging trends in operational designs and patterns. ARIS encompasses research and studies on the general characteristics of revolutionary movements and insurgencies and examines unique adaptations by specific organizations or groups to overcome various environmental and contextual challenges.

The ARIS series follows in the tradition of research conducted by the Special Operations Research Office (SORO) of American University in the 1950s and 1960s, by adding new research to that body of work and in several instances releasing updated editions of original SORO studies.

VOLUMES IN THE ARIS SERIES

Casebook on Insurgency and Revolutionary Warfare, Volume I: 1927–1962 (Rev. Ed.)
Casebook on Insurgency and Revolutionary Warfare, Volume II: 1962–2009
Case Studies in Insurgency and Revolutionary Warfare: Algeria 1954–1962 (pub. 1963)
Case Studies in Insurgency and Revolutionary Warfare—Colombia (1964–2009)
Case Studies in Insurgency and Revolutionary Warfare: Cuba 1953–1959 (pub. 1963)
Case Study in Guerrilla War: Greece During World War II (pub. 1961)
Case Studies in Insurgency and Revolutionary Warfare: Guatemala 1944–1954 (pub. 1964)
Case Studies in Insurgency and Revolutionary Warfare—Palestine Series
Case Studies in Insurgency and Revolutionary Warfare—Sri Lanka (1976–2009)
Unconventional Warfare Case Study: The Relationship between Iran and Lebanese Hizbollah
Unconventional Warfare Case Study: The Rhodesian Insurgency and the Role of External Support: 1961–1979
Human Factors Considerations of Undergrounds in Insurgencies (2nd Ed.)
Irregular Warfare Annotated Bibliography
Legal Implications of the Status of Persons in Resistance
Narratives and Competing Messages
Special Topics in Irregular Warfare: Understanding Resistance
Threshold of Violence
Undergrounds in Insurgent, Revolutionary, and Resistance Warfare (2nd Ed.)

SORO STUDIES

Case Studies in Insurgency and Revolutionary Warfare: Vietnam 1941–1954 (pub. 1964)

TABLE OF CONTENTS

INTRODUCTION...1

PREVIOUS ARIS CONSIDERATION OF
      PHASING CONSTRUCTS.....................................2

OBJECTIVE AND METHODOLOGY.................................9

LITERATURE REVIEW 11
    Process 11
    Results..11
    Early Analysis 12
    Law Literature..12
    Economics Literature 16
    Political Science and Social Movement Theory Literature 21

SYNTHESIS OF PHASING LITERATURE INTO A PROPOSED
      CONSTRUCT 35
    Synthesis into a New Proposed Framework: States of Resistance ..35
        Preliminary State: Incubation 37
        Incipient State: Coalescence.......................................41
        Crisis State: Formalization and Outbreak of Action..................45
        Institutional State: Bureaucratization 49
        Abeyance: Demobilization to Incipience.......................52
        Resolution States...53
            Decline...53
            Radicalization...53
            Institutionalization...54
            Repression 55
            Facilitation 55
        Success..56
            Failure 57
            Co-Optation 57
            Establishment with the Mainstream 58
            Exhaustion..59

ANALYSIS OF ARIS CASE STUDIES AS A
      PROOF OF CONCEPT 59

CONCLUSION.. 74

APPENDIX A. CODED ARIS CASE STUDIES................................ 81

APPENDIX B. ACRONYMS 137

BIBLIOGRAPHY 139

LIST OF ILLUSTRATIONS

Figure 1. Continuum from legal protests to insurgency and belligerency. 13

Figure 2. CIA life cycle of insurgency. 30

Figure 3. Coy and Hedeen's stage model of social movement co-optation. 31

Figure 4. Woods et al. ladder of emotions. 34

Figure 5. Proposed states for phasing construct analysis.......... 36

Figure 6. The concept of relative deprivation, as illustrated by the J-curve 38

Credits:

Figure 1. Continuum from legal protests to insurgency and belligerency. Erin N. Hahn and W. Sam Lauber, *Legal Implications of the Status of Persons in Resistance* (Fort Bragg, NC: United States Army Special Operations Command, 2014).

Figure 2. CIA life cycle of insurgency. US Central Intelligence Agency, *Guide to the Analysis of Insurgency 2012* (Washington, DC: US Government, 2012).

Figure 3. Coy and Hedeen's stage model of social movement co-optation. Patrick G. Coy and Timothy Hedeen, "A Stage Model of Social Movement Co-Optation: Community Mediation in the United States," *The Sociological Quarterly* 46, no. 3 (2005): 405.

Figure 4. Woods et al. ladder of emotions. Michael Woods et al., "'The Country(side) Is Angry': Emotion and Explanation in Protest Mobilization," *Social & Cultural Geography* 13, no. 6 (2012): 567–585.

Figure 6. The concept of relative deprivation, as illustrated by the J-curve. James C. Davies, "Toward a Theory of Revolution," *American Sociological Review* 27, no. 1 (1962): 5–19.

LIST OF TABLES

Table 1. Comparison of Mao translation with US Army ATP 3-05. 3

Table 2. Comparison of Galula's orthodox communist pattern with his bourgeois nationalist shortcut pattern. 4

Table 3. Full list of coded case studies grouped according to their path through the proposed construct. 61

Table 4. Intraelite conspiracy—resolution without crisis. 63

Table 5. Coups and popular revolutions—short crises with decisive resolutions. 63

Table 6. Crisis to institutionalization to resolution. 65

Table 7. Organizational destruction without outright defeat. 69

Table 8. Failed crises followed by repeat attempts. 72

Table A-1. Coded case studies with details. 82

Table A-2. Coded case studies without details. 133

INTRODUCTION

The purpose of the Assessing Revolutionary and Insurgent Strategies (ARIS) project is to produce an academically rigorous yet operationally relevant body of knowledge on insurgency and revolution. This bedrock knowledge includes updated works on underground movements and human factors in the tradition of the Special Operations Research Office (SORO), in-depth historical case studies on contemporary irregular conflicts, and analyses of the legal status of participants in resistance. In this vein, the ARIS team has undertaken the disciplined and methodological study of the phenomenon of resistance, seeking to develop analytical tools to facilitate a deeper and more robust agenda for future research. Included in these tools are the developmental states of resistance presented in this paper, as well as a comprehensive conceptual typology of resistance. Together, these conceptual instruments lay the foundation for a science of resistance that will not only support the needs of the Special Forces community but will also enrich the broader communities of academics and policy makers.

The ARIS *Conceptual Typology of Resistance*[1] unpacks the phenomenon according to its broader attributes in the confines of a formal hierarchy, allowing for the delineation of characteristics in the construction of theories and clarity regarding their applicability. However, a developmental aspect in resistance movements is noted throughout the academic, military, and intelligence literature. Resistance movements are born, grow, mature, escalate, and decline, changing in both shape and character as they progress. For this reason, scholars and analysts have long acknowledged the need to examine these movements not only according to their individual characteristics and behaviors but also according to their phase in development. By conceptualizing and understanding resistance groups and movements according to their development, analysts can then demonstrate patterns in their characteristics and behaviors with greater theoretical clarity.

Numerous developmental theories and frameworks for the examination of resistance movements (and more broadly, for social movements) have been proposed since the 1920s. This paper proposes a developmental framework (or "phasing construct") derived from this large body of literature to facilitate the rigorous study of resistance movements, as well as demonstrates the construct's analytical utility. After reviewing previous ARIS efforts examining the developmental nature of resistance and presenting this work's objective and methodology, this paper reviews key contributions to the concept of phases in resistance in the legal, economic, political science, and sociological disciplines. The totality of the literature is then brought to bear in a new phasing construct that synthesizes commonalities throughout the disciplines into a single framework. The framework is then used to examine a limited sample of historical cases discussed in ARIS studies, with presentation of a coding methodology and limited analysis to clearly demonstrate its utility in the academic study of resistance.

PREVIOUS ARIS CONSIDERATION OF PHASING CONSTRUCTS

The ARIS team previously completed a limited study of phasing constructs.[2] That study focused on the writings of Mao, US Army Field Manual 3-24 (FM 3-24) on counterinsurgency, the US Army Doctrine and Training Publication (ATP) 3-05 on unconventional warfare, David Galula's *Counterinsurgency Warfare: Theory and Practice*, and the 1966 *Human Factors Considerations of Undergrounds in Insurgencies* by the Special Operations Research Office. Before proceeding to how the current study contributes to the understanding of how resistance movements evolve, a review of the analysis and conclusions of this previous work is in order.

The previous ARIS study on contemporary phases of resistance considered first Mao's three phases,[3] characterizing them as follows:

1. Organization, consolidation, and preservation of base areas, usually in difficult and isolated terrain
2. Progressive expansion by terror and attacks on isolated enemy units to obtain arms, supplies, and political support
3. Decision or destruction of the enemy in battle

The US military seemingly adopts this construct in US Army Field Manual 3-24,[4] with the following three phases:

1. Latent and incipient
2. Guerrilla warfare
3. War of movement

According to the field manual, activities in the latent and incipient phase include the emergence of leaders, the creation of an initial organizational infrastructure, training, the acquisition of resources, and engagement in political actions such as protests. The guerrilla warfare phase features small-unit tactics against security forces as well as political actions. During this phase, an insurgency may engage in limited governmental functions in areas under its control. Finally, the war of movement phase features greater military capacity, increased popular support, logistics capability, and territorial control. The field manual distinguishes between two possible forms of success: an insurgency could outright defeat the standing government or force out an occupying power or it could create and maintain a problem that the counterinsurgent is unable to solve definitively, thereby wearing down the counterinsurgent.

A translation of Mao's writings by US Marine Corps Brigadier General Samuel Griffith extends Mao's three phases to seven. Coincidentally the US Army's ATP 3-05[5] also provides a model with seven phases, and the author of the ARIS study suggests that this construct parallels Mao's seven phases (Table 1).

Table 1. Comparison of Mao translation with US Army ATP 3-05.

Mao Translation	US Army ATP 3-05
1. Arousing and organizing the people	1. Preparation
2. Achieving internal unification politically	2. Initial contact
3. Establishing bases	3. Infiltration
4. Equipping forces	4. Organization
5. Recovering national strength	5. Buildup
6. Destroying the enemy's national strength	6. Employment
7. Regaining lost territories	7. Transition

Importantly, ATP 3-05 presents a template for understanding and assisting the planning of an unconventional warfare effort, not a construct for the progression of a resistance movement. Yet, the document does provide some insight into phasing by describing a resistance movement's activities during the unconventional warfare phases. However, those descriptions do not appear until phase four, organization. During that phase, the resistance focuses on establishing a cadre to act as the organizational nucleus of an infrastructure that can withstand the reaction to the group's potential armed activities. In phase five, buildup, the resistance begins to engage in limited offensive operations, but its focus remains on developing an infrastructure to support those limited operations and potentially more extensive operations. Activities apart from operations and support to operations during this phase include providing humanitarian assistance and controlling resources in a way that gains the favor and support of the population. During the employment phase, the resistance movement begins expanded offensive operations, accompanied by efforts to capitalize on successes to build morale and recruitment. These expanded offensive operations consist of a guerrilla warfare and subversion campaign aimed at eroding and disrupting the opponent's morale and resources either as a strategy in itself or in expectation of receiving external assistance. Finally, the ATP 3-05 transition phase presumes the resistance movement's goal is to completely overthrow a standing government or oust an occupying power. Thus, ATP 3-05 assumes the resistance organization will be occupied with the following activities:

- Transforming itself into a functioning government
- Addressing constituent needs
- Demobilizing

The publication does recognize, however, that a movement's goals may be more limited, and a transition phase may not occur at all in such cases.

In *Counterinsurgency Warfare: Theory and Practice*, David Galula proposes a modification to Mao's phasing model based on his in-depth study of the French–Algerian War. Galula adds a so-called bourgeois-nationalist shortcut that entails blind and selective terrorism. Both Galula's shortcut and his formation of the orthodox communist insurgency pattern comprise five steps, of which the first two differ.

Table 2. Comparison of Galula's orthodox communist pattern with his bourgeois nationalist shortcut pattern.

Orthodox Communist Pattern	Bourgeois Nationalist Shortcut Pattern
1. Create a party	1. Blind terrorism
2. Unified front	2. Selective terrorism
3. Guerrilla warfare	3. Guerrilla warfare
4. Movement warfare	4. Movement warfare
5. Annihilation campaign	5. Annihilation campaign

The first two steps of Galula's orthodox communist pattern focus on building a strong grassroots base by using rejected or disenfranchised individuals, primarily intellectuals. The organization created in these steps may be overt or hidden and seeks to garner support from the population by using political actions targeted against the government. The second step, a united front, focuses on maintaining unity during growth and potential support from internal and external allies. It is during guerrilla warfare that both patterns begin with the seizing of power through political acts or armed force. Insurgent activities during this step aim to pull the population into participating or being complicit in the insurgency's campaign. Step four, movement warfare, requires creating and equipping a regular armed force that leverages the intelligence and logistics networks developed during the preceding two steps. Finally, the annihilation campaign entails destroying the counterinsurgent forces and placing the insurgent political party atop the national hierarchy.

The shortcut pattern substitutes the nonviolent building of a core organization and grassroots networks with the building of a military capability. Blind terrorism aims for publicity that attracts support for the insurgency. Selective terrorism focuses on killing local government officials to demonstrate the insurgency's power, alienate the government from the people, and gain the support or complicity of the population. Galula uses the term *shortcut* for this pattern because he believes these first two violent steps build military capacity more quickly and thereby better prepare the insurgency for the subsequent steps of guerrilla warfare, movement warfare, and annihilation campaign.

Galula recognizes that the shortcut pattern risks backlash and loss of support from the population, but he thinks the orthodox communist pattern may risk early defeat in guerrilla warfare because its early stages focus on political activities and building an organization rather than practicing armed violence. However, Galula also highlights the focus in the communist pattern on building a strong political party as a particular strength for postinsurgency governance. He says that even though the shortcut/terroristic approach "may save years of tedious organizational work, . . . the bill is paid at the end with the bitterness bred by terrorism and with the usual post-victory disintegration of a party hastily thrown together."[6] In the broader context, one should recognize that Galula acknowledges these are general patterns that he interprets in history: "While [the patterns] substantially fit the actual events in their broad lines, they may be partially at variance with the history of specific insurgencies."[7] Recall that efforts to identify phasing constructs aim to identify broad patterns, not detailed instructions.

The SORO work is the final publication considered in the previous ARIS work on phases. This publication proposed a series of five phases: organization, covert activity, expansion, militarization, and consolidation. The best explanation of SORO's phasing concept is from the work itself:

> To show how the organizational structure of undergrounds changes in protracted revolutions, it is useful to categorize phases in the evolution of conflict. The first phase is the clandestine organization phase in which the underground begins developing such administrative operations as recruiting, training cadres, infiltrating key government organizations and civil groups, establishing escape-and-evasion nets, soliciting funds, establishing safe areas, and developing external support. During this phase, cell size is kept small and the organization is highly compartmentalized.
>
> The second phase is marked by a subversive and psychological offensive in which the underground employs a variety of techniques of subversion and psychological operations designed to add as many members as possible. Covert underground agents in mass organizations call for demonstrations and, with the aid of agitators, turn peaceful demonstrations into riots. Operational terror cells carry out selective threats and assassinations.
>
> In the third or expansion phase, the organization is further expanded and mass support and involvement are crystallized. Front organizations and auxiliary cells are created to accommodate and screen new members. During the militarization phase, overt guerrilla forces are created. Guerrilla strategy usually follows a three-stage evolution. In the first stage, when guerrillas are considerably outnumbered by security forces, small guerrilla units concentrate on harassment tactics aimed at forcing the government to overextend its

defense activity. The second stage begins when government forces are compelled to defend installations and territory with substantially larger forces. The third stage marks the beginning of the full guerrilla offensive of creating and extending "liberated areas."

During all of these stages, the underground acts as the supply arm of the guerrillas, in addition to carrying out propaganda, terrorist, sabotage, and other subversive activities. Crude factories are set up by the underground and raids are conducted to obtain supplies and weapons. Caches are maintained throughout the country and a transportation system is established. Finances are collected on a national and international basis. Clandestine radio broadcasts, newspapers, and pamphlets carry on the psychological offensive. The underground continues to improve its intelligence and escape-and-evasion nets.

In the fifth phase, the consolidation phase, the underground creates shadow governments. Schools, courts, and other institutions are established to influence men's minds and control their actions, and covert surveillance systems are improved to insure positive control over the populace.[8]

The ARIS work considered the foregoing studies and found limitations in each proposed phasing of resistance. For example, the sentences preceding Mao's seven phases immediately limit their utility by identifying them as necessary to realize that movement's goal of complete emancipation of the Chinese people. Mao's seven stages therefore describe the process that proved successful for that particular resistance movement, and thus they do not necessarily transfer to or accurately describe other resistance movements whose goals may be more limited or whose circumstances are assuredly different. For instance, this construct would fail to describe nonviolent resistance movements, which ostensibly have no need for equipping forces or regaining lost territories. Mao's construct, whether three or seven phases, remains most apt for resistance movements that closely resemble his own. Undoubtedly, this is in part because the construct was created out of that single experience, and the focus was to use force to overthrow the government.

However, it is precisely that background that cabins the construct's utility for describing the trajectory of movements more generally. FM 3-24 specifically considers the phases of an insurgency as opposed to a resistance movement, which can be nonviolent or violent. Additionally, the authors of FM 3-24 understand the work's limitations and state that the construct it proposes simplifies reality for the purpose of assisting readers in understanding and analyzing the phenomenon of insurgency. Moreover, they also recognize that, insofar as the activities of early phases build into later phases instead of ceasing, it is very difficult to determine when an insurgency transitions from phase to phase. Galula's patterns exhibit essentially the same weaknesses as Mao's, namely being constructed out of a single experience (primarily Algeria) and thus not necessarily informative for resistance movements taking place in other circumstances.

By contrast, the US Army wrote ATP 3-05 to guide soldiers in assisting resistance movements, and this document does not suffer the weakness of using only the experience of one specific movement. In that regard, however, some of the phases are not useful for describing the fundamental trajectory of a resistance. The stages of initial contact and infiltration represent required steps for an actor seeking to assist a movement, but these are not necessarily the same as the steps required for a movement to be successful. Yet, ATP 3-05 does provide a useful level of generality that allows it to be mapped to a variety of resistance movements. All movements need to organize, establish sufficient resources, and use those resources. However, ATP 3-05 does not indicate where a threshold might lie for a movement to continue to progress, such that a resistance could cycle through organization, buildup, and employment several times and still be in the same relative position with respect to the government it opposes, to the larger society, and to accomplishing its goal.

This previous ARIS study aimed to examine existing constructs and postulate ways in which they could be improved to better assist those engaged in unconventional warfare to understand how resistance movements evolve. If those conducting unconventional warfare better understand that evolution, they can better assist those resistance movements. Ultimately, that study highlighted the limitations mentioned above. Furthermore, it concluded that what is most needed is detail on the mechanisms[a] and variables that enable a resistance movement to progress or cause it to regress, namely a better understanding of transitions between phases. This might seem to impose a rigid structure on the growth of resistance movements, because the term *transition* can suggest the end of one phase and the beginning of another as though they were clearly demarcated. Instead, a more flexible structure is needed to accommodate the variability observed in resistance movements.

While working on the current effort, the team quickly recognized that there is more literature to be considered than the documents described above. Therefore, this new study widens the aperture of sources considered in a continuing search for a better understanding of how resistance movements evolve and decline. Early results of literature searches in a variety of disciplines revealed that identifying and defining concrete mechanisms and variables that enable movements to progress or cause them to regress remain out of reach. However, it is possible to contribute a synthesized phasing construct that incorporates conceptions of phasing from a broad range of disciplines and sources. Analysts can then use this proposed synthesized construct as a tool to better study and research resistance movements in order to achieve a greater understanding of how they evolve. The disciplines found to contribute the most were, in addition to the doctrinal and historical documents covered above, law, political science, and social movement theory.

OBJECTIVE AND METHODOLOGY

As the ARIS team explored past and current revolutions and resistance movements to identify emerging trends and underlying structures, it discovered that resistance encompasses a broad spectrum of types and manifestations of disruptive movements, and that it is an observable phenomenon with complex and dynamic characteristics and concepts. As mentioned earlier, one of the studies in the ARIS publication *Special Topics in Irregular Warfare: Understanding Insurgency* focused on understanding the phases of contemporary resistance

---

[a] In *Dynamics of Contention*, McAdam, Tarrow, and Tilly define *mechanism* as "a delimited class of events that alter relations among specified sets of elements in identical or closely similar ways over a variety of situations." *Processes*, in turn, are "regular sequences of such mechanisms that produce similar (generally more complex and contingent) transformations of those elements."[9]

movements as described in a few specific documents (e.g., that of Mao Tse-tung). That report, described in more detail in the previous section, concluded that none of these studies adequately covers stages of organizational growth, particularly the mechanisms, determinants, and variables enabling a movement to expand, contract, or stall. The studies mentioned in that report made certain assumptions (e.g., a level of violence) and focused on the actions of successful movements but omitted characteristics and mechanisms that allow a movement to move forward or backward, as well as succeed or fail. Unconventional warfare planners have a need to understand these processes in order to decide, for example, whether it is best to wait and do nothing for now or to take some type of action to support or counter the resistance movement. Without such an understanding, certain actions at certain times may have unintended consequences that should be avoided. Therefore, the present study resulted from the recognition that a more fundamental approach is needed to understand the detailed characteristics and mechanisms that allowed these movements to transition through various stages of organizational growth.

Therefore, this document describes a phasing construct designed as a tool that can be applied to case studies to characterize the organization, infrastructure, resources, leadership, and other factors that allow a resistance movement to change over time, whether it moves forward, moves backward, or stalls. By using this tool to analyze case studies, one might shed light on specific mechanisms that allowed resistance movements to take different paths, including which characteristics, mechanisms, and variables helped or hurt the movement. It is important to emphasize that this tool is not designed to impose conclusions about each phase a priori. Instead, any conclusions should emerge from the research that uses the tool. For that reason, detailed characteristics of each phase are minimal. The methodological development of the proposed phasing of resistance included an interdisciplinary literature review (including the disciplines of law, political science, social movement theory, and economics). The next section describes the process and results of this review.

LITERATURE REVIEW

Process

The team took a multidisciplinary approach, with each team member investigating a particular field of research, including economics, business, political science, social movement theory, law, social psychology, and history. The team used standard academic databases and found that the most salient and productive disciplines were political science and social movement theory. Economics and law publications were found to contain relevant concepts but lacked the depth or breadth that political science and social movement theory literature offered for phasing constructs. The articles found in the economics databases presented no phasing construct as such but instead presumed a conflict between two competing parties. That literature is included here because it could shed light on how entities compete once they reach a certain stage, and in that way the literature may prove useful to the unconventional warfare planner. Law was found to present a phasing construct that will be explained in the section on law literature. However, the legal theory described herein is limited to a progression in the intensity of violence; it does not describe how nonviolent resistance movements progress. History was subsumed by political science. When results were not strictly relevant to the concept of phasing or stages, they were examined for utility in terms of variables that would impact the progression of a resistance movement, the relationships among elements or variables of a movement as they impact a movement's progression, and/or why a field of inquiry would be informative for the study of resistance movements.[b]

Results

The literature review yielded a multitude of phasing constructs and schemas. As mentioned, political science and social movement theory proved to be the most relevant disciplines, so a multidisciplinary presentation was not as useful as it was thought to be at the beginning. However, in this paper, results are organized by discipline to inform

---

[b] This literature review is not comprehensive. Other scholars, such as Maegen Gandy, have conducted research in similar areas.[10]

the reader of what other constructs exist and how those disciplines inform the concept of a movement's progression. Because of the early conclusion that political science and social movement theory offered the most useful contributions, these disciplines will be presented last and constitute the chief contributors to the synthesized construct proposed after the details of the literature review.

Early Analysis

A few themes emanated from this initial research. First, political scientists borrowed from historians, and social movement theorists borrowed from political scientists, illustrating the influence of academics from the early twentieth century on today's theorists. Second, economists use equations to model resistance and insurgency, usually without delineating stages or phases. Those equations could shed light, however, on the relationships between variables impacting the progression of a movement. Third, no constructs exceed five stages or phases. Finally, the literature recognizes porous demarcations between stages and the uniqueness of each movement in its evolution. In this way, the constructs rarely overstated themselves and instead claimed to provide adaptable frameworks useful for analysis but not necessarily for providing information needed for understanding when, where, and how a resistance movement might be influenced to produce a predictable result.

Law Literature

The ARIS study *Legal Implications of the Status of Persons in Resistance*[11] demonstrates that international law maintains a phasing construct for resistance movements in order to determine the applicable law and level of legal protections afforded to participants. The construct is framed within the concept of what constitutes an armed conflict, and it progresses across five stages: (1) the use of legal processes to gain political advantage, (2) the use of illegal political acts, (3) rebellion, (4) insurgency, and (5) belligerency. The first two categories are products of the authors, and the latter three categories come directly from international law. The thresholds for insurgency and belligerency match those for noninternational and international armed conflicts, respectively. Passing these thresholds triggers the application of international humanitarian law at different levels, whereas the stage of rebellion and those preceding it do not trigger any protections beyond constantly applicable international human rights law and domestic legal regimes. A group's progression along these stages is determined by the level of the movement's intensity, duration, and organization. Generally, as those criteria increase, the group progresses from rebellion to belligerency, and vice versa. Some events may defy the idea of progressing from one stage to another, such as the incident adjudicated in *Abella v. Argentina,* after which the Inter-American Court of Human Rights found that the thirty-hour attempted siege of a barracks constituted a noninternational armed conflict even though it did not progress through the stage of rebellion.[12] However, such incidents also serve to illustrate that although this continuum may be clearly demarcated, those clear demarcations do not limit one's analysis of a resistance; one must apply the continuum to the resistance according to the events on the ground.

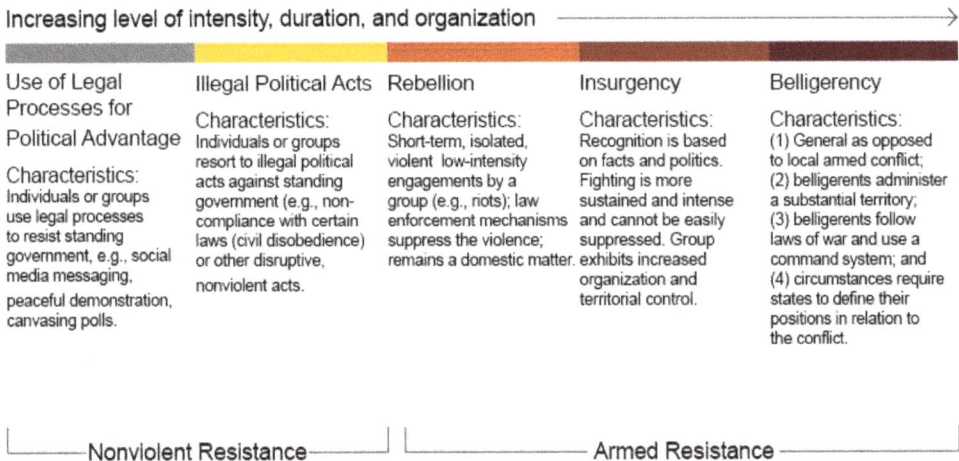

Increasing level of intensity, duration, and organization ————————→

Use of Legal Processes for Political Advantage	Illegal Political Acts	Rebellion	Insurgency	Belligerency
Characteristics: Individuals or groups use legal processes to resist standing government, e.g., social media messaging, peaceful demonstration, canvasing polls.	Characteristics: Individuals or groups resort to illegal political acts against standing government (e.g., non-compliance with certain laws (civil disobedience) or other disruptive, nonviolent acts.	Characteristics: Short-term, isolated, violent low-intensity engagements by a group (e.g., riots); law enforcement mechanisms suppress the violence; remains a domestic matter.	Characteristics: Recognition is based on facts and politics. Fighting is more sustained and intense and cannot be easily suppressed. Group exhibits increased organization and territorial control.	Characteristics: (1) General as opposed to local armed conflict; (2) belligerents administer a substantial territory; (3) belligerents follow laws of war and use a command system; and (4) circumstances require states to define their positions in relation to the conflict.

└————Nonviolent Resistance————┘ └———————— Armed Resistance ————┘

Figure 1. Continuum from legal protests to insurgency and belligerency.

The far left of the continuum includes nonviolent lawful measures, such as permitted protests, litigation, or political campaigns. This category is distinguished from all the others because its actions are within the legal confines that the resisted government put into place. Groups or individuals in this category use the status quo system or systems to effect changes in policy, law, and leadership. However, those in this category rarely, if ever, seek to overthrow the government and system in place.

Because the activities in this category are lawful, the participants are classified as law-abiding citizens exercising their civil and political rights.[13] Resistance in this category looks much like political participation.

Next on the continuum are nonviolent unlawful activities, such as civil disobedience in the form of unpermitted protests and sit-ins or other illegal, disruptive political activities. The host nation's domestic criminal and civil laws apply to these activities and participants, meaning those involved could be charged with criminal or civil offenses but none would be considered prisoners of war, insurgents, or belligerents.[14] Instead, participants would simply be citizens or residents under the jurisdiction of the local law. Apart from standing human rights obligations, no international law is applicable in this category of resistance. It remains a domestic matter for the standing government to address via its law enforcement bodies.[15] For instance, participants in the US civil rights movement who engaged in sit-ins or other civil disobedience were arrested for violating local laws, and the government was bound only by its human rights commitments, not by international humanitarian law. Equally, participants in the South African antiapartheid resistance who undertook illegal activities were subject to South African law and protected only by that law and its compliance with human rights commitments undertaken by South Africa. Resistance in this category looks like civil disobedience, and the law treats it as a matter regulated by local laws.

Activities in these first two levels share the characteristics of being nonviolent and being able to be carried out by individuals acting independently. When resistance activities become violent, international law classifies them under one of three possible levels depending on the intensity of the violence, the duration of the resistance, and the organization of the resisting group.[16] If rebellion, insurgency, and belligerency are all insurrection, then rebellion is juvenile, insurgency adolescent, and belligerency adult insurrection, each with increasing violence, duration, and organization.

Resistance in the category of rebellion features low-intensity, isolated, short-term violence executed by groups that may be organized but not to a high degree.[17] Additionally, the groups' violent activities, such as riots or uncoordinated small-arms attacks, do not seriously challenge the standing government's monopoly on force. Instead, law enforcement entities are capable of containing and suppressing the resistance. Just as with nonviolent levels of resistance, the applicable law is the host nation's criminal and civil law. Again, the resistance remains a matter of domestic laws and law enforcement; international humanitarian law does not yet apply.[18] The acts of rebellion are simply crimes, and the participants in rebellion are criminals.[19]

As the intensity of the violence increases, the resistance persists despite the government's efforts to subdue it, and, as the resistance's organization increases, it can transition from rebellion to insurgency. An insurgency is characterized by more sustained and intense fighting that the government is unable to suppress easily. To be classified as an insurgency, a resistance exhibits increased but not extensive organization as well as control over territory.[20] These factors also align with the standard for finding that a noninternational armed conflict exists, meaning what was previously a group of criminals perpetrating violent acts against the government now constitutes a nonstate armed group engaging the standing government in an armed conflict. Consequently, the international humanitarian law protections that accompany a noninternational armed conflict also apply.[21]

Additionally, once a situation becomes an armed conflict, then it begins to trigger additional rights, duties, and obligations of the standing government and outside parties. For instance, outside nations owe neutrality to the host nation, so as not to lend legitimacy to the insurgency by recognizing it either as a viable contender for status as lawful government or, more dramatically, recognizing it as the lawful government of the host nation.[22] This demonstrates that, with regard to law, resistance centers on the group's goal of becoming the government of the country. The group can accomplish this goal by asserting control over the people through a monopoly on force, having a high degree of organization, and surviving the standing government's attempts at suppression.

On the far right of the continuum is belligerency, which requires (1) an armed conflict of a general as opposed to local character; (2) belligerent occupation and administration of a substantial portion of national territory; (3) that the belligerents conduct hostilities according to the rules of war under a responsible command authority; and (4) circumstances that make it necessary for outside states to define their attitudes by recognizing the belligerency.[23] Essentially, belligerency looks like a full-fledged civil war during which both sides, of relatively equal effectiveness, contend for the right to be the lawful government. Prominent examples include the American Revolution and the US Civil War, with some nuanced caveats. Crucially, a belligerency possesses enough organizational and military capabilities that it rivals the state and controls and administers extensive territory. When a resistance has become a belligerency, the armed conflict has grown to resemble a state-on-state armed conflict sufficient to merit treating it as an international armed conflict under international humanitarian law.[24] Consequently, participants in a resistance that is legally classified as a belligerency are combatants and therefore receive amnesty for their wartime acts and prisoner of war status if they are captured. This classification also means, however, that the resistance owes the same duties and protections to the state's armed forces, such as to capture when possible and to provide the rights afforded prisoners of war.[25]

Law thus provides a relatively clear phasing construct with three identified variables, namely intensity, duration, and organization, that determine where along the continuum a resistance movement resides. However, this construct has limited utility for analyzing the progression of nonviolent resistance movements. It features only two phases addressing nonviolent resistance and draws only one distinction between them, the legality of the activity. An additional limitation is in the continuum's purpose. The continuum was not derived for studying resistance movements but instead was designed to help determine when an armed conflict exists in order to know which legal protections apply to the relevant actors. The legal continuum nonetheless sheds light on how resistance movements evolve and provides a useful way to analyze at least one category of resistance movements (i.e., armed), if not all resistance movements.

Economics Literature

Focusing on economic factors in insurgency progression, the team identified more than fifty papers by searching literature in these databases: ABI/INFORM Complete, Business Source Complete, EBSCOhost, EconLit, Economist Intelligence Unit, LexisNexis Academic, and Stratfor Global Intelligence. Of the economics papers found during the literature search, Naylor's work provides the most useful economic insights into the developmental stages of resistance.[26] The paper analyzes the finances of a wide variety of insurgent groups over several decades. Naylor notes that modern underground politics and modern underground economics share the perception that the current formal state apparatus is not legitimate. Resistance movements have financial obligations (i.e., expenditure responsibilities) in order to meet their political responsibilities.

Naylor presents the results of his analysis as an evolution through three stages of financing characterized by changes in expenditure responsibilities and fund-raising activities:

1. **Zones of contention:** This is considered the earliest stage, where the rebel organization engages in sporadic hit-and-run operations (e.g., kidnapping or assassination). The targets include individual symbols of the state, such as government officials, local police stations, and isolated military outposts. The expenditures are relatively small and overwhelmingly military and logistical. The primary form of resistance fund-raising is called predatory, which includes maritime fraud (which is much broader than piracy), counterfeiting, bank robbery, and ransom kidnapping.

2. **Zones of exclusion:** During this stage, the resistance organization becomes more deeply entrenched in society or in particular geographic regions. The organization now openly disputes the political power of the state by conducting low-intensity warfare against the infrastructure of the formal economy. The main targets of attacks cease being only small political symbols and become more economic. The objective is not yet the capture of territory but is instead the destruction of basic infrastructure, industry, and commerce to cause investment to shrink, capital to flee, production to fall, unemployment to rise, and inflation to increase. The rebel organization's expenditures have grown larger to include an increasing social security provision for the dependents of the militants as well as provision of some assistance to the population whose support they are attempting to gain. Therefore, fund-raising needs to yield a steady and reliable income at the expense of the legal, formal economy. The primary form of the rebels' fund-raising is called parasitical and includes more organized activity such as embezzlement, protection rackets, illegal gambling, and illegal drug distribution (similar to organized crime). Naylor describes this as a "revolutionary taxation" of income and wealth.

3. **Zones of control:** At this stage, the rebel organization becomes more secure in its exclusive hold on territory. Its expenditures include military and social service expenses, as well as capital expenditure on the provision of infrastructure and the development of an economy that parallels the official economy. Sources of revenue now come from indirect taxation, including sales taxes on domestic commerce, import/export taxes on foreign trade, and "user fees" for public services. The primary form of rebel fund-raising is called symbiotic, which means that the rebels' business approximates mainstream business and the goal is the provision of goods and services (both legal and illegal) to the population within the territory it now controls. Even though fund-raising activities are increasingly overt and legitimate looking, asset management becomes more difficult, and activities require the guerrilla group to interface with the formal and international economy similar to the way in which so-called white-collar criminals hide and launder their returns.

Naylor mentions that these stages may sometimes overlap. Importantly, he also makes the point that insurgent groups are not simply extensions of the criminal economy but are motivated by political advantage rather than profit. That is, they use their income to further their political goals. This underground economy may provide a means of production and distribution of goods and services that would be considered legal if it operated inside the formal economy, but the state is unable to exercise regulatory control or taxation.

Although resistance groups engage in fund-raising activities inside the country, external (outside the country) funding offers two advantages: (1) funding is likely to be regular and consistent as long as the resistance group is seen to be implementing the political objectives of the external sponsor; and (2) outside aid often solves some of the insurgent forces' logistical problems by supplying them with heavy and sophisticated weapons. The main disadvantage of external funding is that, if aid from outside sponsors were revealed, the resistance movement might appear to be the tool of a foreign government, which could result in a reduction in domestic support because of the perceived foreign intrusion. Naylor classifies external financial support as private or public. Private outside sponsors may be motivated by sympathy to the goals of the insurgency, but their fund-raising may skirt legality, such as when charitable relief money is diverted into black market arms purchases. External private fund-raising may also be involuntary, such as when insurgent groups extort money from well-to-do emigrant communities. Naylor mentions that public funding, meaning funding coming from a government source, is usually laundered through front organizations so that the sponsoring state can maintain plausible deniability.

The other economics papers found during the literature review did not describe individual states of resistance but presented mathematical models that include some potentially useful variables.

Berman, Shapiro, and Felter[27] used a mathematical model and game theory to analyze resistance as a struggle over information among government, rebel, and community (noncombatant) players. They found that the players might choose an equilibrium state (i.e., Nash's equilibrium) even though it is not a global optimum for them. The authors note that their result has broad implications:

> Noncombatants are not enfranchised and the government puts no weight on their welfare, yet they receive services in equilibrium anyway. This service-provision effect is common to [that described by] Akerlof and Yellen (1994), and [the] U.S. Army (2007). It results from the optimal behavior of a government trying to motivate information sharing by noncombatants as a means of suppressing violence.

That is, by providing economic aid and services despite rebel activities, the government can contribute to the popular perception that it is capable of maintaining law and order so that the population is more likely to share information with the government. This information sharing then helps constrain rebel violence.

Guttman and Reuveny[28] developed a game theory model and applied it to autocratic regimes only. One interesting conclusion of their model is that policies that expand economic ties to autocratic regimes appear to make them even more totalitarian.

As mentioned earlier, only the Naylor paper discusses different states of resistance, albeit in strictly economic terms. However, the economics papers generally provide some potentially useful variables that might influence insurgency and counterinsurgency strategies even when they do not clearly define separates states of resistance:

- Type of fund-raising activities of the insurgents (predatory, parasitic, or symbiotic)
- Government spending on reconstruction and infrastructure (small projects are better than large projects because there is more effective oversight and less corruption)
- Degree of the general population's satisfaction with the level of goods and services provided by the government
- Situational awareness of the government versus that of the insurgents
- Violence intensity ratio between the insurgents and the government
- Effectiveness of the insurgents' coercion of the general population
- Targeting accuracy of the violence both by the insurgents and the government
- Sensitivity of the population manifested by the way in which they remember and perceive violent events that directly or indirectly affect them
- Effectiveness of government repression on the insurgents versus government infrastructure investment
- Number of attacks by insurgents
- Combat effectiveness of the government versus the insurgents
- Recruitment rates for the insurgent versus the government
- Number of insurgent leaders
- Number of insurgent foot soldiers

It should be noted that Naylor's categories of resistance fund-raising tend to be more aligned with violent insurgencies. In particular, Naylor's predatory and parasitical fund-raising activities are associated with the threat of violence, if not overt acts of violence. In contrast, non-violent insurgencies could include voluntary fund-raising from sympathetic individuals or organizations. Naylor's symbiotic fund-raising could include such voluntary efforts.

Finally, integrating Naylor's zones into the proposed construct would mean assuming resistance groups predominantly practice certain types of fund-raising within a given state, so there would need to be an a priori assignment of fund-raising type to the different phases of the construct. Such assumptions are not considered appropriate here because the tool was designed to avoid imposing a priori conclusions about each phase.

Political Science and Social Movement Theory Literature

The comprehensive literature search suggests that political science and social movement theory contribute the most relevant information to the study of phases of insurgency and resistance. Of the numerous references the team found, the following references were deemed the most relevant to phasing constructs. Some of these date from the early twentieth century, but they serve as the foundations for later studies. Because recent studies heavily reference these earlier studies and because of their perceived value by the academic community, we believe that these earlier studies remain relevant and provide significant information.

In his seminal 1927 book *The Natural History of Revolution*, Lyford P. Edwards lays the intellectual foundation for later theorists on the phasing of resistance movements, arguing that revolutions are an extreme symptom and result of long-understated social change, rather than a cause of social change themselves. The stage theory Edwards proposed posits that (1) preliminary and (2) advanced symptoms eventually lead to (3) an outbreak of revolution, which then escalates to (4) crisis before an eventual fatigued return to (5) normality. Preliminary symptoms of revolution are characterized by "an increase in general unrest . . . at first very vague and indefinite,"[29] during which stage "it is probable that nobody has the remotest notion of any revolution."[30] The numerous advanced symptoms include "the transfer of the allegiance of the intellectuals"[31] and publicists to the cause, the realization of economic incentives for revolution, and the growth of a social myth justifying resistance.[32] "The outbreak of revolution," argues Edwards, "is commonly signaled by some act, insignificant in itself, which precipitates a separation of the repressors and their followers from the repressed and their followers."[33] This results in the initiation of the revolution because the preparation for revolution in the crowd psychology has come to fruition. The potential crisis stage of revolutions comes with the rise of radicals who enforce a reign of terror. The success of moderates or conservatives to secure power usually progresses directly to a return to normalcy, a process that either way comes with a sense of slow deflation and exhaustion.

Crane Brinton's 1938 *The Anatomy of Revolution* was an early and influential work in the growing body of literature theorizing on the developmental nature of revolutions and the characteristics associated with each point of progression.[34] Referred to as "commonalities" among revolutions, Brinton arrives at his stages and related characteristics through the comparative study of the English (1642–1660), American (1775–1783), French (1789–1799), and Russian (1917) Revolutions. Brinton asserts that revolutions have four stages in common: preliminary (the old order), first (moderate regime), crisis (radical regime), and recovery (Thermidorian reaction).[35]

The preliminary-stage symptoms of revolution under the old order, according to Brinton, are preceded by the development of a middle class. Members of this middle class then perceive injustice in their economic position and form into "cells" that garner support from intellectuals and meet obstacles to translating their developed economic influence into political participation. Additionally, the government itself is revealed to be financially crippled and inept, and foreign nations may seek to aid the opposition to weaken the government. When these factors coalesce, according to Brinton, revolution may emerge.[36]

The first-stage symptoms of revolution include the emergence of concrete actions against unpopular policies, the formation of clear competing groups among the opposition (moderate and radical), and the emergence of a small and active minority that represents the grievances of the majority. While this vocal minority has not yet organized into centralized planning, the government is nevertheless eventually forced to attempt to repress the building insurrection, but it fails. In this way, the standing regime is shown to be unable to rule. Events in this stage can include financial breakdowns, symbolic actions, and dramatic events. The first stage of revolution ends after the moderate opposition forces seize power and establish their legitimacy.[37]

The moderate and radical revolutionary factions clash in the second stage of revolution. Because they compromise some revolutionary goals, the ruling moderates earn the ire of both the conservative supporters of the old regime and their rivals while simultaneously providing freedom of speech and other rights. This enables the radicals to stage a coup against the new regime. The extremists then typically succeed because of their organized and disciplined fanaticism, and they proceed to concentrate power in the hands of a strongman and implement a period of terror to maintain control and enforce conformity with the gospel of the revolution. This radical habit for violence increases the chances for the reemergence of either foreign or civil war.[38]

Finally, the recovery stage is characterized by ebb in the fervor of revolutionary feeling and a resistance against the radical revolutionary regime (Thermidorian reaction), and as life slowly returns to normal, a ruler comes to power and revives an adjusted manifestation of the original regime. The new regime's repression of radicals and forgiveness of moderates is accompanied by an aggressive revival of nationalism and the revival of earlier social mores and norms (religious, social, etc.).[39]

In his article "Sequence in Revolution," Paul Meadows conducted a literature review and attempted to integrate various stage theories on the nature of revolutions. He eventually proposed a three-phased structure that traces revolutionary development with the "attitudinal evolution in revolution" in mind, where there are two simultaneous developments toward achieving one thing while combating another. First, incubation is a precrisis stage during which the revolutionary movement develops unrest and a frame of reference through the subtle spread of new ideas, attitudes, and values that eventually lead to critical self-awareness of something to be combated and something to be achieved. The movement simultaneously develops a sense of insecurity (that something is wrong) alongside weakening loyalties to the state. Second, action is the crisis phase during which the revolutionary group transitions from academic to martial values in organized groups, adopting a structuralized form of protest and extroverted tendencies to promote strategic values, eventually pushing to remove obstructive social conditions. Likewise, the movement transitions from characterizing the state's abuse as systemic injustice, attempting to exert new social control through techniques that manipulate social goods, symbols, violence, and practices. Finally, adaptation is the postcrisis phase during which the revolution must consolidate authority and structuralize stability through cathartic and constitutional means.[40]

Rex D. Hopper's article "The Revolutionary Process: A Frame of Reference for the Study of Revolutionary Movements"[41] put forward a new and deeply influential phasing hypothesis that stands today as formative in social movement theory, which only slightly adds to or revises

Hopper's framework since its publication in 1950.[42] Hopper's process is claimed to be a synthesis of previous works, including those of Sorokin, Edwards, and Brinton, which resulted in four stages of revolution: the preliminary stage, the popular stage, the formal stage, and the institutional stage. The author then describes each stage of revolution in terms of its characteristic conditions, typical processes, effective mechanisms, types of leaders, and dominant social forms.

According to Hopper, the characteristic conditions present in "the preliminary stage of mass (individual) excitement and unrest" include (1) "general restlessness"; (2) "the development of class antagonisms"; (3) "marked governmental inefficiency"; (4) "reform efforts on the part of the government"; (5) a "cultural drift in the direction of revolutionary change"; and (6) the evident "spread and socialization of restlessness." The typical social process during the preliminary stage of revolution is "milling," or "circular interaction," in which unorganized restlessness emerges from unknown causes, with "uncertainty in reference to the ends toward which action should be directed." The mechanisms that are effective in influencing people in this stage of unrest include "such devices as agitation, suggestion, imitation, propaganda, et cetera," which suits the emergence of a leader in the form of an agitator "who stirs the people not by what he does, but by what he says." Finally, the dominant social form in the preliminary stage of revolution is a "psychological mass" composed of anonymous people "from all walks and levels of life . . . responding to common influences but unknown to each other . . . [having] little or no organization on the level of mass behavior."[43]

Second, "the popular stage of crowd (collective) excitement and unrest" is "a time of the popularization of unrest and discontent; a time when the dissatisfaction of the people results in the development of collective excitement . . . [where] individuals participating in the mass behavior of the preceding stage become aware of each other." The characteristic social conditions of this stage include (1) "the spread of discontent and the contagious extension of . . . unrest and discontent," expressed in "increased activity, growing focus of attention, and heightened state of expectancy"; (2) "the transfer of allegiance of the intellectuals" to the aggrieved population, including the "identification of a guilty group . . . [and the] development of an 'oppression psychosis'"; (3) "the fabrication of a social myth," including "collective illusions, myths, and doctrines, . . . the economic incentive to revolutionary action, . . . [and] a tentative object of loyalty"; (4) "the emergence of conflict with the out-group and the resultant increase in in-group consciousness"; (5) discontented organization "for the purpose of remedying the threatened or actual breakdown of government"; and (6) "the presentation of revolutionary demands which if granted would amount to the abdication of those in power." In this context, the typical processes of social contagion and collective excitement give rise to an "effort to develop esprit de corps" by emerging leaders, typically prophets (who offer "a new message and a new philosophy of life") and reformers (who attack "specific evils and develops a clearly defined program"). Finally, the dominant social form of the popular stage of revolution is "crowd formation" from the mass of the previous stage, where a "psychological crowd" transitions into an "acting crowd."[44]

Third, in the formal stage of "the formulation of issues and formation of publics . . . the movement must strike deeper than sensationalism, sentimentalism, fashion, and fad. It must come to appeal to the essential desires of the people." This stage is characterized by two conditions: (1) "the fixation of motives (attitudes) and the definite formulation of aims (values)"; and (2) "the development of an organizational structure with leaders, a program, doctrines, and traditions." The former condition is reached through the factionalization and internal conflict within the revolutionary movement, and the latter is accompanied by the increasing breakdown of government authority, the development of dual sovereignty or a provisional government, precipitating factors for the seizure of power by radical factions, and a "lull" between the seizure of power by radicals and the use of terror "as a control technique." The processes of revolution in this stage are "(1) discussion and deliberation, (2) formulation, and (3) formalization." The dominant mechanisms of this stage develop "group morale and ideology" by instilling conviction in the purpose of the revolution, in the belief that the goals will be realized, and in the belief that the purpose represents "a sacred responsibility which must be fulfilled." The group ideology is usually established through "(1) a statement of objectives, purposes, and premises; (2) a body of criticism and condemnation of the existing social order . . . ; (3) a body of defense doctrine to justify movement; (4) a body of belief dealing with policies, tactics, and practical operations; and (5) the myths of the movement." Leadership emerges in this stage of revolution, and the dominant social form of a public takes hold, "marked by the presence of the discussion of, and a collective opinion about, an issue."[45]

Finally, during "the institutional stage of legalization and societal organization," "the out group must finally be able to legalize or organize their power" toward established authority. The sociopsychological conditions characteristic of these actions are classified as causal (transitional) and resultant (accommodative). Causal characteristics include psychological exhaustion, moral letdown, and economic distress, all of which undermine the foundations of the revolution. Resultant characteristics are numerous, including an end to the use of terror, increasing centralization of power, social reconstruction, dilution of the revolutionary ideal, and the institutionalization of the movement as a permanent organization "that is acceptable to the current mores." Processes in this stage depend on discussion and deliberation "for fixing policies and determining action" toward the institutionalization of the movement, a point in development on which "the success of the entire revolutionary movement" depends. The effective mechanisms for this stage "are well-nigh innumerable," perfecting the tactics established in the earlier stages of the movement. The emergent leaders required for the movement in this stage are the administrator-executives, who "deliberately employ all [the] various types of leadership" established in the earlier stages. Finally, the dominant social form of the institutional stage is the shift of the public into a society, acquiring "organization and form, a body of customs and traditions, established leadership, and enduring division of labor, social rules and social values."[46]

In his contribution to Jo Freeman and Victoria Johnson's *Waves of Protest: Social Movements since the Sixties*, Frederick D. Miller "examines factors that contribute to the decline of social movements and organizations that comprise them."[47] Miller's "model of movement decline" contends that the history of movements are influenced by (1) "events in the world that influence the availability of resources and the success of tactics; (2) movement ideologies that influence tactical and structural choices; and (3) movement organizational structure, which also influences tactics and ways of accessing and mobilizing resources."[48] Although these factors are closely related and "cannot be studied independently," Miller nevertheless contends that "the history of any movement organization is determined by an interaction between [these] factors." Additionally, an organization can decline while the movement itself continues, as long as other movement organizations persist.[49]

According to Miller, social movements and movement organizations can decline through "repression, co-optation, success, and failure":[50]

- "Repression occurs when agents of social control use force to prevent movement organizations from functioning or prevent people from joining movement organizations. The variety of repressive tactics includes indicting activists on criminal charges, using infiltrators to spy on or disrupt groups, physically attacking members and offices, harassing members and potential recruits by threatening their access to jobs and schools, spreading false information about groups and people, and anything else that makes it more difficult for the movement to put its views before relevant audiences."[51]

- "Co-optation strategies are brought into play when individual movement leaders are offered rewards that advance them as individuals while ignoring the collective goals of the movement. Such rewards serve to identify the interests of those co-opted with those of the dominant society. . . . [Co-optation] is most likely to be effective with movements of powerless constituencies who have few skilled activists."[52]

- "Success [is] a bit more complicated. . . . It is conceivable that a movement could set goals, accomplish them, and subside, with success obviating the need for the movement. This is rare, however, probably limited to instances where people organize solely to achieve one goal. . . . Few movements see the satisfaction of all their demands. Instead, they make or are forced into compromises that only sometimes are advantageous to the movement. . . . In obtaining concessions from the dominant system, movement organizations often have to relinquish some portion of their claim to represent an independent radical opposition. This process of absorption brings social movement organizations into the structure of interests in the polity, converting them into interest groups." Success can also harm movement organizations by exposing them to internal rifts as they grow. In the case of new movement organizations, seeking association may sap resources.[53]

- "Failure at the organizational level takes two major forms: factionalism and encapsulation. Factionalism arises from the inability of the organization's members to agree over the best direction to take. . . . Encapsulation occurs when the movement organization develops an ideology or structure that interferes with efforts to recruit members or raise demands." Although neither of these organizational-level failures necessarily means a given movement will decline, the movement will decline if no other organizations persist to prolong it.[54]

In the revised 2011 edition of his book *Power in Movement: Social Movements and Contentious Politics*, Sidney G. Tarrow proposes a "mechanism-and-process approach"[55] to examining cycles of contentious politics. He identifies several mechanisms as "combined in complex cycles of contention": dispositional mechanisms ("such as the perception and attribution of opportunity or threat"), environmental mechanisms ("such as population growth or resource depletion"), and relational mechanisms ("such as the brokerage of a coalition among actors with no previous contact by a third actor who has contact with both").[56] Tarrow then outlines the mechanisms for contentious mobilization shared by "challengers and those they face": the "interpretation of what is happening" (frame the field of contention); the perception of opportunities and threats; and the creation or appropriation of resources to take advantage of opportunities and ward off threats. "Challengers engage in innovative collective action" while those they face "organize to oppose or appease them."[57] Mechanisms for demobilization include repression, facilitation, exhaustion, radicalization, and institutionalization.[58] Processes of diffusion ("when groups make gains that invite others to seek similar outcomes") include direct (relational), indirect, and mediated diffusion.[59] The mechanisms and processes constituting a "Great Event" that triggers cycles of contention include making opportunities, innovating in the repertoire (tactics), waging protest campaigns, and forming coalitions.[60]

The *Guide to the Analysis of Insurgency*, originally published in 2009 and updated in 2012 by the US Central Intelligence Agency (CIA), presents "the life cycle of an insurgency and keys to analysis" as part of "an analytic framework designed to assist in evaluating an insurgency." The cycle includes four stages: preinsurgency, incipient conflict (noted as early [growth] in the figure), open insurgency (noted as middle [mature] in the figure), and resolution (end). Despite these "common stages of development," the guide insists that the manifestation of these stages and their characteristics are not universal and are case specific: "some will skip stages, others will revisit earlier stages, and some will die out before reaching the later stages" (see Figure 2). Although characteristics "are identified for each stage," they are all nevertheless "continuous and cumulative."[61]

First, the preinsurgency stage, which is "difficult to detect," is primarily composed of underground activities. The insurgency "has yet to make its presence felt through the use of violence," and "actions conducted in the open can easily be dismissed as nonviolent political activity." Organization, leadership, grievances, and group identity in this stage are only emerging and beginning to develop, as are tasks such as recruitment, training, and the stockpiling of arms and supplies. Noted keys for analysis include the preexisting historical, societal, political and economic conditions; identified and publicized grievances; the distinguishing group identity; "the first signs of insurgent recruitment and training [that] may emerge;" the initial gathering of arms and supplies; and the government's reaction, which is "perhaps the most important determinant of whether a movement will develop into an insurgency."[62]

Third, during the open insurgency stage, "no doubt exists that the government is facing an insurgency." The insurgent forces are openly "challenging state authority and attempting to exert control over territory," with more frequent attacks that have "probably become more aggressive, violent, and sophisticated." The role of any external support for the insurgency also becomes more apparent. The political and military factors of the insurgency (at this point attempting to replace, rather than merely undermine, state authority), as well as any external assistance, are noted as particular areas for analysis in this stage.[64]

Finally, the resolution stage marks the theoretically inevitable conclusion of an insurgency, resulting in one of three end states: the insurgent victory, negotiated settlement, or government victory. An insurgent victory "is the only potential outcome that is likely to be clear-cut," but it may mark the beginning of a new conflict. Negotiated settlements, however, will likely "have many false starts, delays in implementation, and attempts by spoilers to undermine the agreement," as well as a persistent risk of renewed violence. A government victory is likewise drawn out, "marked by gradual decline in violence as the insurgents lose military capabilities, external assistance, and popular support" as limited violence persists to an indistinct end.[65]

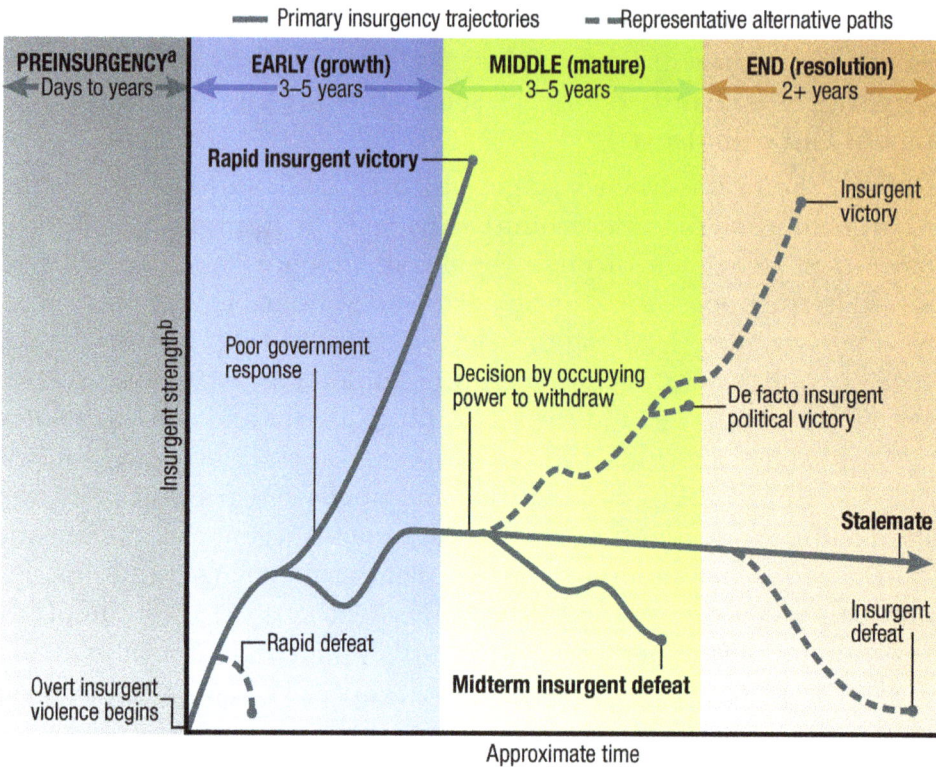

— Primary insurgency trajectories   — — Representative alternative paths

| PREINSURGENCY[a] ←Days to years→ | EARLY (growth) ←3–5 years→ | MIDDLE (mature) ←3–5 years→ | END (resolution) ←2+ years→ |

Insurgent strength[b]

Rapid insurgent victory

Insurgent victory

Poor government response

Decision by occupying power to withdraw

De facto insurgent political victory

Stalemate

Rapid defeat

Insurgent defeat

Overt insurgent violence begins

Midterm insurgent defeat

Approximate time

[a] Preinsurgency activities include the emergence of insurgent leadership, creation of initial organizational infrastructure and possibly training, acquisition of resources, and unarmed political actions, such as organizing protests.

[b] Insurgent strength is a subjective measure of the size of a movement as well as its ability to mount attacks and inflict casualties, popular support, logistics capacity, and/or territorial control. The insurgency trajectory will vary according to insurgent and government actions.

[c] The decision by an occupying power to withdraw is commonly made four to seven years into the conflict.

Figure 2. CIA life cycle of insurgency.

In their 1960 article, scholars from the University of California, Los Angeles, added to the literature on directions of decline and factors contributing to the failure of social movements, particularly in those movements that did not move past their incipiency. The authors contribute four avenues, particularized to mass-based movements that were notably successful at their outset (according to the case examined), by which young movements can fail. First, movements can fail by neglecting to establish "a preexisting network of communication linking those groups of citizens most likely to support the movement." A second avenue is the "failure of an emergent leader to incorporate . . . [other] leaders into his organization." Third, the young movement may lack "a program to which a major section of the [participants] could give whole-hearted support." Finally, failures may become "highly publicized" and "conspicuous," creating a fatally "weakened… public image."[66]

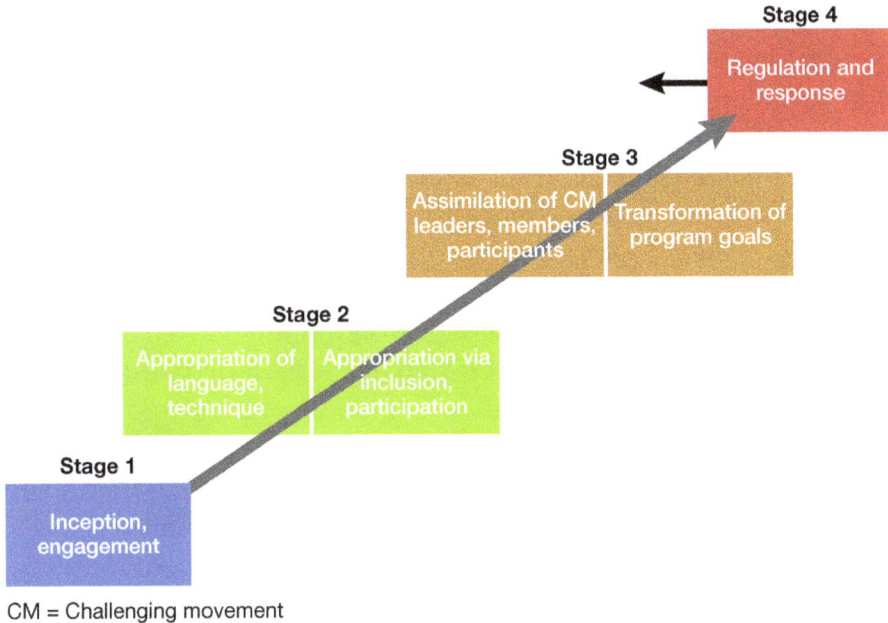

CM = Challenging movement

Figure 3. Coy and Hedeen's stage model of social movement co-optation.

Building on Frederick D. Miller's contributions on the decline of social movements, Patrick G. Coy and Timothy Hedeen contributed further depth to the literature with their article, "A Stage Model of Social Movement Co-Optation." Outlining four stages (see Figure 3), Coy and Hedeen "analyze the evolution of community mediation and identify and degrees and dimensions of [challenging movement] co-optation."[67]

The first stage is the assumed state of the movement arising "in response to a set of grievances or unfulfilled needs" in "a segment of the population."[68] This stage of inception and engagement may include demands for change, the establishment of alternative institutions, and the state's or vested interests' recognition that there is a need for reform.[69] The second stage consists of two steps of appropriation. First (stage 2a), the state or vested interests appropriate the challenging movement's language and methods by dismissing the challenging movement's values or redefining its terms as antithetical to those values. Second (stage 2b), the state or vested interests challenges the movement's leadership and power base by allowing modest participation in policy making and/or implementation, creating the perception of power sharing and potential institutionalization, both of which divert the challenging movement's energies away from directly challenging the state or its vested interests.[70] The third stage also consists of two steps. In the first step (stage 3a), the leaders of the challenging movement assimilate into the state or its vested interests through employment opportunities and/or the state's or vested interests' development of a controlled alternative to the challenging movement. In the second step (stage 3b), the state or vested interests then develop institutions to support affiliated alternatives to the challenging movement, setting priorities and changing goals in accordance with their own interests, which in turn force the challenging movement to restructure according to the states' or vested interests' goals.[71] The fourth and final stage of social movement co-optation is regulation and response, during which the state (or vested interests) "routinizes, standardizes, [and] legislates" any resulting changes, and expectations are shaped to align with state or vested interests. The challenging movement then defensively responds by developing institutions to support, maintain, buffer, insulate, or protect its own goals.[72]

In his 1964 article "New Theoretical Frameworks," Ralph Turner discusses collective behavior and conflict and offers three possible constructs for social movements.[73] The goal of his paper is to deconstruct prior frameworks and suggest new, more complex frameworks of movements. The discussion highlights three existing constructs of collective behavior and conflict and attempts to modify each with further research:

1. **Process resolution versus unfolding.** This construct explores the life-cycle approach to a movement, where correlates for attributes of collective behavior are different at different stages of a movement. Processes are often contradictory rather than part of a single, locked stage structure, and nuance comes from analyzing the relative strength and direction of each process. This method presents a fixed goal and sequence and all development is measured against that goal. The output thus artificially limits the complexity of the development. The matured version of this construct (the

unfolding) introduces variables that are not specifically tied to any outcome but rather can be observed at any stage.

2. **Imminent versus interactive determination.** Public labeling of a movement impacts the character, recruitment, ideology, portrayal, and strategy of the movement. For example, how the public observes the movement determines the way "in which members think of the movement and themselves, the type of ideology they develop for the movement, and the aspects of the ideology and value which become most salient."[74] Four types of movements are classified: revolutionary, peculiar, respectable-factional, and respectable-nonfactional. "The character of the movement is not intrinsic and is never fixed, but is always a product of the visible actions of the movement, the public response to those actions, and movement adaptations to this response."[75]

3. **Emergent norm versus contagion theory.** The contagion approach combined the conclusions of LeBon, Freud, and Trotter in discussing nonrational processes. The approach of the alternative process is in the characterization of an "overpowering impression of homogeneity but that more careful observation always shows the unanimity to be an illusion." Six significant characterizations of crowds stemming from the emergent norm theory are illustrated.[76]

In "The Country(side) Is Angry," Michael Woods and his fellow authors explore the significance and role of the activists' emotions and the way in which these emotions evolve.[77] Previously, when specifically connected to protest mobilization, emotions had been disregarded or, at best, deemed insignificant in scholarly research. Woods et al. propose a ladder of emotions and explore how space plays into the emotional aspect of the politics of protest, asserting that emotional attachment to a place can stimulate political mobilization.

Key principles underlying the study of emotions and protests have established the framework for multidisciplinary investigation of emotions in social movements. However, most studies explore specific aspects of the following principles:

- Emotions that are most relevant to political behavior are strongly connected to value systems.
- Emotions are collective as well as individual.

- Emotions are crucial to the formation and mobilization of social movements.

- Social movements are active in transforming and reproducing emotions.

- Emotions can distort the actions and strategies of social movements.

"Few, if any studies, however, have followed the emotional journeys of participants through protest activity, interrogating the changing emotions of participants at different stages of mobilization."[78]

Figure 4 illustrates the Woods et al. ladder of emotions. Protest mobilization increases as the numbers progress from 1 through 5. At 6, emotions cause a downward trend in mobilization.

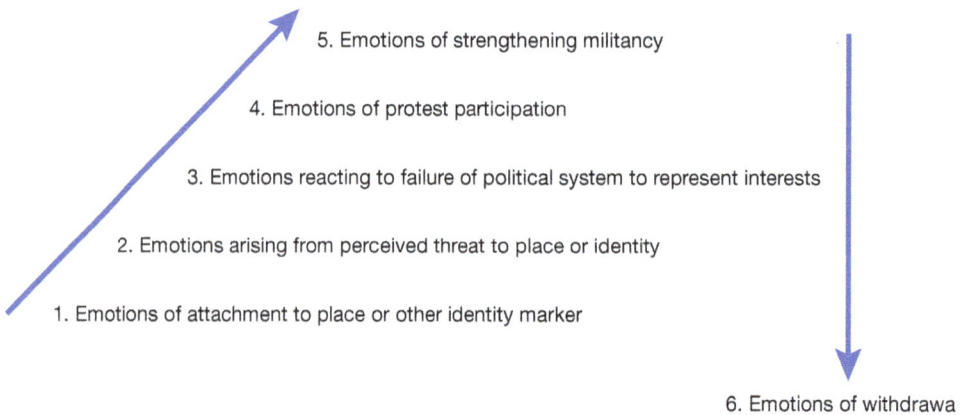

5. Emotions of strengthening militancy

4. Emotions of protest participation

3. Emotions reacting to failure of political system to represent interests

2. Emotions arising from perceived threat to place or identity

1. Emotions of attachment to place or other identity marker

6. Emotions of withdrawa

Figure 4. Woods et al. ladder of emotions.

Woods et al. explore how progressing from one rung to the next involves the translation of one set of emotions into a different set of emotions. Emphasizing emotions places a burden on the observer to pinpoint potential irrational emotions and convert them into rational, progressive actions. The authors concede that the ladder is a mere prototype; most emotions will not progress to the point of protest mobilization. Nonetheless, Woods et al. perceive that a gap in existing research was created because scholars emphasized rational action when considering protest mobilization; they attempt fill that gap by providing a different perspective.

Building on the review of literature on law, economics, political science, and social movement theory for phasing constructs, the next section synthesizes the phasing constructs featured in the literature. Not all sources are incorporated, and some are incorporated more heavily than others are. For instance, the proposed construct strongly features the work of early twentieth-century historians later echoed by later-twentieth-century social movement scholars. This is in part because the proposed synthesis must be flexible enough to apply to resistance movements generally, rather than to a particular type of resistance. The constructs covered in the previous ARIS study on phases are limited because of their focus on armed, violent, military-based resistance movements. It is important to note that this proposed synthesis is intended to be a tool used for further study, allowing researchers to explore finer details of the mechanisms and variables involved in resistance movements' progressions and regressions.

SYNTHESIS OF PHASING LITERATURE INTO A PROPOSED CONSTRUCT

Synthesis into a New Proposed Framework: States of Resistance

The phased framework of a resistance movement's life cycle proposed in this paper is a synthesis of the multidisciplinary literature on the subject. Drawing from commonalities in the literature and the evolution of academic thought, as well as from military theories and doctrine, this adapted model works in conjunction with the ARIS conceptual typology. Together, these tools facilitate the intensive and detailed study of resistance cases to both broaden and deepen robust institutional and scholarly knowledge of resistance as a societal phenomenon. This phasing construct is a tool that users can apply in comparative case studies through data set development and coding, case selection, research design, and even less robust contemporary case studies, allowing the user to shed light on specific mechanisms that allowed resistance movements to take different developmental paths. It is important to emphasize that this tool seeks to avoid imposing conclusions from the conventional wisdom that a few conspicuous factors define transition in resistance (e.g., use of violence or holding territory). Instead, any conclusions and developments should emerge from deeper research examining the shared dynamics of violent and nonviolent resistance movements.

After the detailed presentation of this new model and its influences alongside historical examples, all forty-six case studies in both volumes of the *ARIS Casebook on Insurgency and Revolutionary Warfare* will be coded according to their developmental paths. The coding methodology and comparative analysis will demonstrate a proof of concept for using the proposed phasing in future research on resistance.

The proposed five states of resistance (Figure 5) are consistent with many of those proposed in the literature: preliminary, incipient, crisis, institutionalization, and resolution. The first four states are theorized to be consecutive, although a resistance can revert to a previous state. A resistance can move into a resolution state from any of the other four, but the particular type of resolution will vary, and some resolutions are particular to some states.

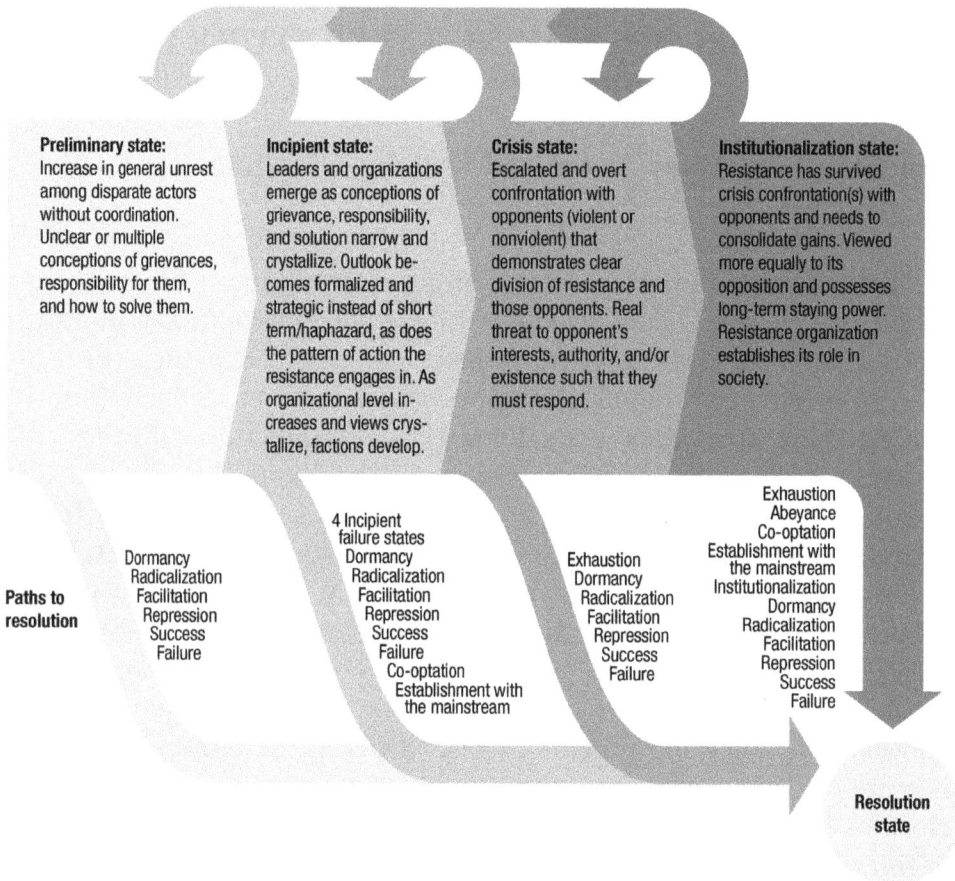

Preliminary state: Increase in general unrest among disparate actors without coordination. Unclear or multiple conceptions of grievances, responsibility for them, and how to solve them.

Incipient state: Leaders and organizations emerge as conceptions of grievance, responsibility, and solution narrow and crystallize. Outlook becomes formalized and strategic instead of short term/haphazard, as does the pattern of action the resistance engages in. As organizational level increases and views crystallize, factions develop.

Crisis state: Escalated and overt confrontation with opponents (violent or nonviolent) that demonstrates clear division of resistance and those opponents. Real threat to opponent's interests, authority, and/or existence such that they must respond.

Institutionalization state: Resistance has survived crisis confrontation(s) with opponents and needs to consolidate gains. Viewed more equally to its opposition and possesses long-term staying power. Resistance organization establishes its role in society.

Paths to resolution

Dormancy
Radicalization
Facilitation
Repression
Success
Failure

4 Incipient failure states
Dormancy
Radicalization
Facilitation
Repression
Success
Failure
Co-optation
Establishment with the mainstream

Exhaustion
Dormancy
Radicalization
Facilitation
Repression
Success
Failure

Exhaustion
Abeyance
Co-optation
Establishment with the mainstream
Institutionalization
Dormancy
Radicalization
Facilitation
Repression
Success
Failure

Resolution state

Figure 5. Proposed states for phasing construct analysis.

Preliminary State: Incubation

The first state of resistance is the preliminary state, also referred to as "latent" in Army doctrine[79] or "emergence" in modern social movement theory.[80] The preliminary state's most defining feature is the manifestation of unorganized and unattributed unrest—unorganized because actors are unconnected, and unattributed because the unrest lacks a common narrative about the source of the problem. First proposed by Edwards, this is the infancy of resistance, featuring only "an increase in general unrest" with "very vague and indefinite" conceptions or not even "the remotest notion" of organized movement.[81] Edwards describes how a certain amount of restlessness is normal and healthy, but this restlessness and unease tend to increase when the population is unable to satisfy its elemental desires for security and recognition.

Another preliminary symptom of revolution is what Edwards calls a "balked disposition." This occurs when the population begins to perceive that its legitimate aspirations and ideals are being repressed or hindered, albeit without knowing exactly how or why. This discontent with the established routine of life eventually becomes contagious so that individual unrest becomes social (i.e., collective) unrest. Continued repression of this social unrest leads to the population's mutual response of sympathy, which Edwards called "rapport." Individuals are no longer reluctant to express discontent publicly, and those freely expressing their discontent recognize others similarly expressing discontent. Individuals reinforce each other's discontent so that grievances echo across society and thereby continue to heightened tension and unrest. However, Edwards emphasizes that there is no thought of organization or revolution at this stage.

Brinton likewise theorizes that this earliest state is fundamentally distinguished by the presence of disparate factors that only begin to coalesce into resistance in the next state.[82] Brinton and Meadows both characterize the preliminary state as a process of incubation, during which the resistance movement is born from the combination of general unrest with a sense of insecurity.[83] Hopper spoke to these dynamics in more detail as "milling" or "circular interaction," where unorganized restlessness emerges with "uncertainty in reference to the ends toward which action should be directed."[84]

These descriptions of general unrest leading to sustained resistance may be enhanced by using the example of relative deprivation, a concept described by Davies,[85] Gurr,[86] and others. Davies[87] presented a J-curve graph as a way to illustrate this concept (see Figure 6, in which the curve appears as an upside-down J). It represents the "intolerable gap between what people want and what they get"; perceptions, founded or unfounded, that a population is owed more than it receives, whether in tangible goods or intangible opportunity, can breed vague, general discontent.

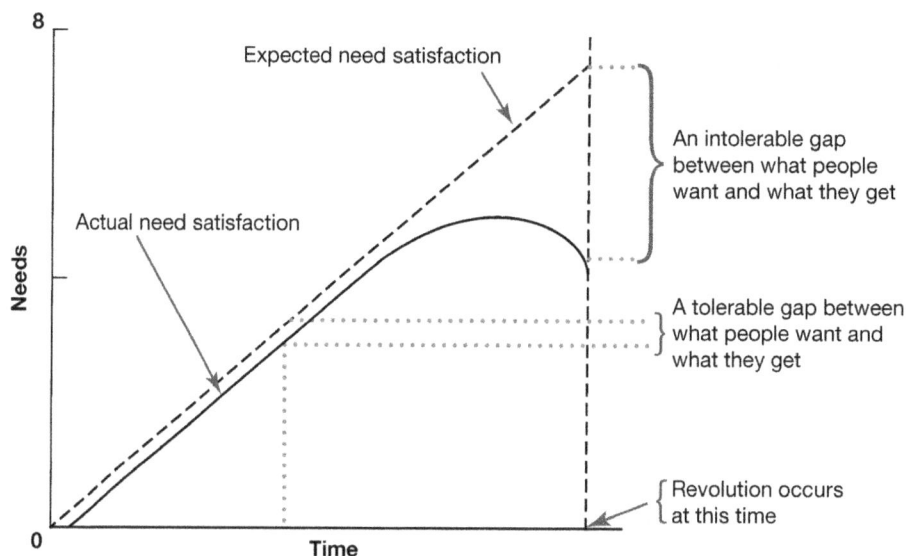

Figure 6. The concept of relative deprivation, as illustrated by the J-curve. This is an example of one way to specify the type of general unrest characteristic of the preliminary state.

Pettigrew[88] reviewed the previous works of Davies and Gurr but attributed the original concept of relative deprivation to Stouffer and defined this concept as "a judgment that one or one's ingroup is disadvantaged compared to a relevant referent and that this judgment invokes feelings of anger, resentment, and entitlement." Pettigrew concluded that there are three psychological processes involved in relative deprivation: (1) cognitive comparisons; (2) cognitive appraisals (that these individuals or their ingroups are disadvantaged); and (3) conclusion that these perceived disadvantages are unfair and angrily resented. Pettigrew notes that although relative deprivation has been criticized as being too psychological and thus more relevant to individuals than groups, he believes that the concept can contribute as a sociological predictor of revolution. To support this proposition, he cites the hypothesis of Williams,[89] which stated that the role of relative deprivation in sustained protest could be enhanced when a large cohesive group who recently achieved significant economic and political power suddenly perceives relative deprivation and when the control elements of society are widely perceived as weak, indecisive, and disunited. Furthermore, Pettigrew[90] suggested that there is evidence that group relative deprivation could be assessed by some type of cross-sectional interview survey of the population and that such a survey may have predictive value for revolutions.

Note that Mao, ATP 3-05, Galula, and SORO did not propose phases of resistance analogous to this preliminary phase; those constructs assume the preexistence of an aggrieved population and offer the "organization" of these premotivated individuals as the first phase.[91] The CIA's *Guide to the Analysis of Insurgency* is similarly limited in its "left-of-boom" perspective by presenting preinsurgency as the first stage, during which an insurgency is organizing but "has yet to make its presence felt through the use of violence." Nevertheless, the guide does allude to the importance of this coalescing stage and process by calling out notable keys for analysis, including preexisting historical, societal, political, and economic conditions as well as vocalized grievances and group identities.[92]

Examining the historical cases of Solidarity in Poland, the Taliban in Afghanistan, and the Tamil Tigers in Sri Lanka brings greater clarity to the shape and contours of the preliminary state of resistance. Uncoordinated action by disparate groups, aggravating factors, and the gradual or rapid development of narrative frames for collective mobilization are common themes. Later states, on the other hand, are coordinated by, through, and among movement groups and leaders, all exercising matured strategies for collective action, clearer goals, and discernible narratives for why others should participate in or support the resistance.

Solidarity

The years 1956 to 1976 mark the preliminary state of the Solidarity movement in Poland. In this state, the movement was not represented through the single organization of Solidarity but rather through emerging disparate groups and activity. Aggrieved social groups, primarily students, workers, and intellectuals, remained disparate and acted independently. For example, students did not participate in workers' strikes in 1956 and 1970, and the workers did not participate in the student protests in 1968. Meanwhile, intellectuals distanced themselves from protests in favor of focusing on concessions and reform within the government. Despite separate, uncoordinated actions, demands for free organization, speech, and association were largely in concert across the resistance movement's groups. During this period, economic downturns and subsequent government cuts brought about an increase in general unrest and insecurity. Despite a hike in unrest, the goals and strategies of the resistance remained vague and uncertain.

This state was also characterized by a renewed focus on and vocalization of historical, political, and religious grievances among Polish citizens. A history of Russian oppression tracing back to the eighteenth century blended with dissatisfaction against the standing Soviet-backed regime to foster a Polish identity for the resistance against a common enemy. The influence of the church in the resistance also contributed to the narrative of a common struggle against oppression by providing the resistance with symbols and rituals that resonated with the people. This renewed focus on history and identity provided a salient narrative to a population willing to come together in struggle against a common enemy, enabling Solidarity to amass a popular following that reached fourteen million members.

Taliban

From 1990 to 1994, the Taliban resistance in Afghanistan was in the preliminary state. Although the group remained in this stage for less time than most movements do, like the preliminary states of other resistance movements, the Taliban's preliminary state stemmed from the combination of immediate and historical factors. The Taliban emerged after decades of perceived religious, ethnic, and political oppression at the hands of the government. In this case, oppression from a Soviet-backed regime in the 1970s, Soviet occupation in the 1980s, and ultimately oppression at the hands of the Mujahidin produced widespread grievances. The Afghan Mujahidin resistance against Soviet occupation in Afghanistan from 1978 to 1990 introduced general concepts of Islamic revolution and political rule. The Taliban co-opted these concepts to provide ideological foundations (Islamic fundamentalism), clear objectives (implementation of sharia law, political Islam, expulsion of foreign invaders), and justifications for resistance (law and order, religious morality) against the corrupt Mujahidin and later American invasions.

With previous resistance against the Soviet Union and preexisting conditions providing a foundation, high levels of insecurity, violence, and fragmentation brought about a new resistance against the Mujahidin in the preliminary state of the Taliban. A decentralized and fractionalized Mujahidin authority created disparate regional power holders that developed contrasting visions for political and economic solutions, most notably the more moderate traditionalist and more radical fundamentalist factions. High levels of political instability resulted from this fragmentation and the Mujahidin's inability to consolidate power, producing lawlessness and the breakdown of government functions. This political vacuum allowed for the expansion of drug production, primarily to fund regional authorities, resulting in economic dependence on poppy cultivation, high levels of unchecked criminal activity and corruption, and increased economic disparity between power holders and civilians. In addition to political instability and economic disparity, the absence of a common, non-Islamic enemy allowed internal ethnic divisions to come to the forefront, most importantly between the Pashtun population and the Tajik Mujahidin leadership. The Taliban organized in 1994 as a Pashtun-led offshoot of the fundamentalist faction of the Mujahidin, marking transition into the incipient state, and was quickly able to gain territory and draw support by using these political, economic, and ethnic grievances that intensified in the preliminary state.

The Liberation Tigers of Tamil Eelam

The preliminary state of the Tamil resistance in Sri Lanka began with Sri Lankan independence from British rule in 1948 and continued until the creation of the Liberation Tigers of Tamil Eelam (LTTE) in 1976. Newfound independence engendered tension between ethnic Sinhalese and Tamil populations. This tension stemmed from Britain's past favoritism of the Tamil population and marginalization of the majority Sinhalese population. After a brief period of cooperation after Sri Lanka gained independence, the ethnic divide increased and became politicized when a Sinhalese government was elected in 1956 on a nationalist "Sinhalese-Only" platform. By 1958, general unrest emerged on both sides, and Tamil-led antidiscrimination protests were answered by violent anti-Tamil riots. All government attempts to incorporate Tamil demands were met with virulent popular opposition, and in 1972, the government incorporated anti-Tamil policies into the constitution, solidifying the position of a politically, economically, and socially marginalized Tamil population.

By the mid-1970s, the Tamil resistance's pattern of activities turned from nonparticipation and noncooperation with the government to armed insurrection. Despite this generalized shift in the Tamil resistance toward armed conflict, numerous disparate organizations emerged and competed for Tamil popular support. Violence ensued between organizations seeking to represent the Tamil cause, including the LTTE predecessor, the Tamil New Tigers, producing competing and unclear ideas about the goals of Tamil resistance as well as doubt and uncertainty about the resistance movement's longevity. By 1976, the LTTE emerged as the primary organization for Tamil resistance with the goal of Tamil secession and independence, marking the movement's transition into the incipient state.

Incipient State: Coalescence

Transition to the next state of resistance occurs when disparate factors coalesce into discernible collective action, loose or formal organizations mobilize, and participants have a clear sense of what is wrong and who is responsible.[93] This phase is called coalescence in much of social movement theory,[94] but it has also been referred to in the literature as the incipient phase.[95] The defining feature of the incipient state is the development of intentional organization and a common narrative. The movement begins to take shape, either formally or informally, with the emergence of leaders who seek "to develop esprit de corps" and enact "crowd formation" from thinking in kind to acting in concert with a shared purpose.[96]

Hopper distinguishes this phase as when "discontent is no longer covert, endemic, and esoteric; it becomes overt, epidemic, and exoteric," meaning discontent changes from hidden, localized, and internalized to explicit, widespread, and shared, as "individuals participating in the mass behavior of the preceding stage become aware of each other."[97] This level of active contention mirrors Pettigrew's third psychological process of relative deprivation, during which participants perceive that their disadvantages are unfair and react angrily. In other words, individual discontent evolves into a collective movement as the discontented, disconnected groups and individuals become self-aware as a new and distinct group that must communicate and organize to act in concert. Beyond the emergence of leadership, coordination among once disparate actors causes the conception of "strategies for success" to take shape, making the movement "more than just random upset individuals; at this point they are now organized and strategic in their outlook."[98]

Edwards first characterized this stage as "advanced symptoms" of revolution, including the movement's development of an intellectual cadre and a social myth justifying resistance.[99] Brinton likewise characterizes this phase (which he refers to as first-stage symptoms) as involving the emergence of concrete action for the cause, and he highlights the early signs of factions within the resistance movement.[100] Because actors once disparate during the preliminary state act together to develop strategies to achieve the movement's goals in the incipient state, participants are likely to disagree, which may lead to internal factions. Meadows spoke of both incubation and coalescence in a single phase, but the characteristics of the latter are evident in his description of how the movement develops an ideational frame of reference (i.e., specific goals) for something to be combated and something to be achieved.[101] Moving from an incipient to crisis state occurs when the movement has grown powerful enough to pose a serious threat to its opponent, culminating in significant escalation in its actions and initial confrontation with the government. An incipient movement may also reach a state of resolution instead of progressing, but the ways in which a movement reaches resolution from the incipient state are unique from the ways in which movements can resolve from other states in the construct.

During the incipient state, the Maoist and derivative phasing constructs begin. In Mao's *Guerrilla Warfare*, incipience is evident in his organization and political unification phases. The movement's activities in both of these phases began to emerge in the incipient state. Similarly, SORO's organization and covert activity phases most clearly apply to the incipient state, as well as expansion activities, which presumably manifest in later incipience (after organization and strategic formation have taken place).[102] Although the CIA's *Guide to the Analysis of Insurgency* speaks exclusively to violent militant resistance, it nevertheless applies to incipience of general resistance as well, focusing on the emergence of a movement's identity, leadership, theory of victory, popular support, and logistical concerns.[103] The guide insists that a movement "enters the incipient . . . stage when the insurgents begin to use violence,"[104] despite the fact that movements may abstain from insurgent or violent strategies through incipiency and even into some following stages. However, this is largely because the guide considers insurgent conflicts, not the more general phenomena of resistance that allows for nonviolent movements, which may or may not elect to use violence. Nevertheless, the general dynamics of a resistance movement's incipiency reinforce the proposed characterization.

The Iranian Revolution, Shining Path in Peru, and Ukraine's Orange Revolution each provide illuminating examples of the incipient state of resistance, during which actors gain momentum through developed and continually refined strategies, repertoires for action (i.e., tactics for acts of resistance), organization, leadership, narratives, and recruitment. These efforts build the strength, reach, influence, and capabilities of the movement. If successful in the incipient state, the resistance will then be strong or influential enough to pose a real or existential threat to its opponents, thus transitioning to an escalated confrontation in the crisis state.

Iranian Revolution

The period from 1977 to December 1978 in the Iranian Revolution exemplifies the incipient state because revolutionary leaders emerged, the opposition forces coalesced, discernible collective action took place, and a clear sense of what is wrong and who is to blame developed. Ayatollah Khomeini, though in exile, solidified his role as the leader of the revolution and successfully united a previously fragmented opposition, including merchants (*bazaaris*), urbanites, workers, intellectuals, and clergy. Islamic religious scholars (the ulema), by this time largely aligned with Khomeini, developed as an intellectual cadre for the revolution and established a religious narrative justifying revolution against the secularizing and repressive regime. Critical political actors joined the call for revolution during this time. In early 1977, the opposition National Front Party distributed open letters accusing the shah's regime of corruption and repression. In response, a January 1978 government-backed newspaper article denounced Khomeini and incited widespread popular protests. In late 1978, Khomeini met with National Front officials in Paris, signaling an official unification of critical opposition actors.

Meanwhile, the resistance showed discernible collective action and undertook a pattern of activities, including strikes, rallies, and protests, that are recognizable as perpetrated by the resistance in opposition to the shah's regime. Public marches and forty-day commemorations of killed protesters illustrated swelling popular support and mobilization for the revolutionary cause. Massive workers' and teachers' strikes throughout the fall of 1978 crippled the regime. As the shah's regime continued to weaken, revolutionary forces clarified demands for an Islamic republic. By December 1978, the shah agreed to reorganize the government to appease the opposition but was swiftly forced out of Iran, signaling the onset of the crisis state.

Shining Path

The Peruvian communist insurgency Shining Path was in the incipient state from 1968 to 1980. The organization's transition into this state was marked by its founding under the leadership of Abimael Guzman, a professor of philosophy at the University of San Cristóbal de Huamanga in Ayacucho. Guzman established the Communist Party of Peru in 1968, but by 1970, his disenchantment with the party's unwillingness to take up arms led him to form Shining Path. Guzman's extensive university and community networks in the Ayacucho region coalesced to fill the ranks of Shining Path and spread the call for revolution. During this period, the organization focused on strategic expansion into indigenous communities in the highlands region, where economic depression and long-term racial tensions provided a willing audience for Shining Path's Maoist doctrine. Shining Path members traveled to and lived in rural communities to learn the culture, spread the communist ideology, and recruit villagers. The incorporation of popular support from the highland communities into the activist-based organization in Ayacucho signaled the coalescence of disparate social groups in opposition to the Peruvian government for the shared purpose of communist revolution.

In addition to coalescence and the spread of the insurgency, internal organizational developments marked this period. Shining Path refined its Maoist ideology and strategic goals; established a narrative of class and racial oppression; and developed a membership indoctrination process, recruitment criteria, and a leadership structure. During this state, Guzman articulated the organization's strategic goals in a five-point program: (1) convert the backward areas into advanced and solid bases of revolutionary support; (2) attack the symbols of the bourgeois state; (3) generalize violence and develop a guerrilla war; (4) conquer territory and expand the bases of support; and (5) lay siege to the cities and bring about the total collapse of the state. In line with the group's focus on the indigenous highland communities, a narrative of "pure native peasant communism" solidified during this time; this narrative later became a weakness as the insurgency tried to expand into urban areas. The group developed recruitment and training tactics, including reference-only admittance into the organization and a two-year training and indoctrination program that culminated with an initiation ritual during which the new member was required to "cross a river of blood" through murder to prove loyalty and commitment to the cause, in one case causing the deaths of two French tourists. In the spring of 1980, Guzman declared the start of armed struggle and attacks against the government began, marking transition into the insurgency's crisis state.

Orange Revolution

The Orange Revolution in Ukraine was in the incipient state from 1999 to November 2004. Opposition to President Kuchma's standing government coalesced throughout 1999–2000, most notably after three events in 2000: Kuchma's rumored authorization of the murder of an investigative journalist, his removal of Deputy Prime Minister Yulia Tymoshenko, and the subsequent removal of his popular prime minister, Leonid Yushchenko. These events, on top of claims of electoral fraud in the 1999 election, brought previously disparate groups together and provided a clear sense of what is wrong and who is to blame. Discernible collective action and mobilization against the regime ensued, exemplified by the "Ukraine without Kuchma" campaign and antigovernment protests in Kiev throughout 2000 and 2001. This period also saw the development of resistance leaders and an intellectual cadre. Interestingly, this leadership evolved largely from Kuchma's removed inner circle, most notably Tymoshenko and Yushchenko. The group developed strategies and a pattern of action with a focus on resistance and revolution through electoral channels and peaceful protest. One example is Yushchenko's formation of the "Our Ukraine" Party in 2002 and the party's get-out-the-vote campaign leading up to the 2004 elections. Believing that the overwhelming public support for new leadership (Kuchma, facing term limits, handpicked Yanukovich to run as his successor) would bring about change simply by getting people to the ballot box, the resistance focused solely on getting citizens to vote rather than advocating for its own nominee, Yushchenko, against Kuchma's handpicked successor, Yanukovich. The first round of elections in October 2004 did not produce a winner, and a runoff election was planned for November 21. Rampant electoral fraud in the runoff election incited massive protests, marking the revolution's transition into the crisis state.

Crisis State: Formalization and Outbreak of Action

The crisis state is the phase that distinguishes resistance movements from social movements more generally. Although all resistance movements can be considered social movements, not all social movements are

45

resistance movements. The essential characteristic defining a resistance in the crisis state is escalated confrontation with opponents, constituting a decisive moment (however long or short) of culmination for the movement. As an incipient resistance continues to gain power and influence through the formalization of organization and efforts, its opponents are thus relatively more vulnerable, incentivizing an outbreak of escalation in the resistance movement's action (nonviolent, violent, or both) that brings about a state of heightened confrontation and real risk to the opponents of the resistance. The idea of resistance or revolutionary movements naturally moving toward and culminating in a state of outbreak and crisis is prominently acknowledged in the early literature on revolutions (including the works of Edwards, Brinton, and Meadows).[105] This idea disappears only after the study of revolutionary movements progressively generalizes past resistance and toward the study of social movements as a whole.[106]

Edwards captured the crisis state in his characterization of "the outbreak of revolution," which "is commonly signaled by some act, insignificant in itself, which precipitates a [decisive] separation of" the resistance from its opponents,[107] bringing about more formalized resistance that poses a significant threat. However, Edwards's model differs from the proposed construct presented in this paper, characterizing the "crisis" itself as the potential eventuality of violent clash between resistance factions, rather than an existential threat and heightened confrontation between the resistance and its opponents (which, in Edwards's view, occurs during the outbreak stage). Brinton likewise encompasses the state of crisis within an earlier stage, considering a revolution's outbreak and seizure of power first-stage symptoms. However, many of the events Brinton characterizes in the first stage are those that may be considered crisis-state scenarios (decisive loss of legitimacy, financial collapse, strong symbolic actions, dramatic events, and others).[108] Meadows highlights the crisis phase as the central peak in revolutions, characterized by a movement's shift from academic to martial values (if such a transition has not yet occurred), structuralized collective action, a push for the removal of obstacles to its cause, and the strategic exertion of new power and control over social goods.[109]

Hopper's "formal stage" likewise contains aspects of the crisis state, including the decisive breakdown of the opponent's authority, emerging perceptions of dual sovereignty or provisional authority, and factors precipitating resistance factions gaining power. Hopper offers numerous factors indicating the resistance movement's maturation and formalization, and these factors are also critical in the crisis state, including "the development of an organizational structure with leaders, a program, doctrines, and traditions," all serving to deepen "group morale and ideology" through various means.[110] Hopper further cemented "the formal stage" into social movement theory thought, describing it as when "the movement must strike deeper than sensationalism, sentimentalism, fashion, and fad. It must come to appeal to the essential desires of the people."[111]

Convention in modern social movement theory does not formally isolate a crisis state before bureaucratization, because such escalations of contention are particular to resistance.[112] Maoist and derivative constructs also straddle the crisis state between transitional stages, particularly buildup and employment (as labeled in ATP 3-05) or SORO's transition from expansion to militarization.[113] The CIA guide does not include a state of crisis, as violent insurrection is assumed. However, the "open insurgency stage" characterizes the crisis well. In this stage, the resistance is "overtly challenging" and "displacing" the authority of its opponents, attempting to "exert control" over social or political goods in its purview (in insurgency, this is most visibly evident in territory).[114]

Illustrative examples of the crisis state can be found in the Chechen revolution in Russia, the Provisional Irish Republican Army (PIRA), and the Viet Cong. In each, the ongoing resistance comes to a culminating point of heightened confrontation when there is a decisive schism in the actions of resistance participants and opponents, whether those actions are political or violent. It is important to note that although the cases recounted below are all violent resistance movements, nonviolent resistance movements likewise experience a crisis state. A clear example is the Egyptian Revolution of 2011, during which mostly nonviolent measures escalated to a crisis state via decisive confrontations in Tahrir Square. When the Mubarak government's attempts at violent suppression failed, and the military abandoned Mubarak by refusing to fire on crowds, the movement succeeded in ousting the authoritarian president.[115]

Chechen Revolution

The crisis state of the Chechen Revolution lasted from December 1994 to August 1996. This is a relatively straightforward example of resistance in the crisis state, marked by decisive separation of resistance and opponents, high vulnerability

47

of the resistance movement's opponents, perceptions of dual sovereignty, and a clear escalation in the resistance movement's activity and confrontation with opponents. The opponents weakened after the fall of the Soviet Union in 1991 and subsequent regional crises. The declaration of an independent Republic of Chechnya in 1992, along with Russian preoccupation with other regional independence movements it considered higher risk (e.g., Lithuania, Georgia, and Ukraine), allowed Chechnya to operate as a de facto state. By 1994, a perception of dual sovereignty existed and Russia was vulnerable to the demands of the resistance. December 1994 marks a clear outbreak of action and heightened confrontation as Russian troops entered the Chechen capital, Grozny. War between Russia and Chechnya, known as the First Chechen War, ensued until a cease-fire in August 1996. The cease-fire agreement marks the resistance movement's transition into the institutional state.

Provisional Irish Republican Army

From January to July 1972, the PIRA was in the crisis state. The maintenance of barricaded "no-go" and "free" zones in Derry and Belfast during this period contributed to perceptions of the movement's provisional authority and separation of the resistance from its opponents. There was heightened contention, and the resistance movement escalated action after British troops killed thirteen civilian demonstrators in what is known as Bloody Sunday. Public and international backlash against the British government increased its vulnerability to the resistance movement's demands.

Additionally, the resistance escalated its actions in retaliation, most notably via a violent bombing campaign. With the intensified threat from the resistance, the British government initiated secret talks between PIRA and the British secretary of state. The secret talks were unsuccessful, and in July 1972 PIRA bombs exploded across Belfast in what is known as Bloody Friday, resulting in nine civilian deaths. Despite backlash after the Bloody Friday bombings, the PIRA persisted through the crisis state and transitioned into the institutional state.

Viet Cong

The Viet Cong is an example of a resistance movement that experienced the crisis state twice. Its first crisis state was from 1959 to 1964, and the second from January 1968 to March 1968. North Vietnamese leader Ho Chi Minh's 1959 announcement of armed revolution against South Vietnam signaled the Viet Cong's transition from the incipient state to the crisis state. This announcement decisively separated the resistance from its opponents (South Vietnam and the United States). Additionally, North Vietnam's creation of a Central Office of South Vietnam to oversee Viet Cong operations and a political arm for the Viet Cong, the National Liberation Front (NLF), signaled the formalization of the resistance as willing and able to exert its power and control. Finally, a coup in December 1963 demonstrated the vulnerability of the South Vietnamese regime. Taking advantage of this vulnerability, the Viet Cong intensified attacks and achieved its first military victory against US forces, signaling its transition into the institutional state. The Viet Cong resistance slid backward from the institutional state to the crisis state with the onset of the Tet Offensive in January 1968. This act signified an escalation in the resistance movement's activity and brought about heightened military confrontation between the Viet Cong and US forces. The Viet Cong slid further into the incipient state with the breakdown of the Tet Offensive in March 1968. Other North Vietnamese resistance groups, primarily the Peoples' Army of Vietnam, replaced the Viet Cong as military representatives of the resistance and carried on the conflict with US forces until Vietnam was reunified in July 1976.

Institutional State: Bureaucratization

Referred to as bureaucratization in modern social movement theory,[116] the institutional state of resistance exists if the group or movement either persists through or gains strength from the crisis state of confrontation with its opponents, deepening its organizational and strategic prowess as an equal opposition player with broadened appeal and long-term staying power. In other words, the essential characteristic of a resistance in the institutional state is an established role in society. Edwards and Brinton do not address the institutional state, largely because of the cases considered in their studies (Enlightenment and Soviet revolutions).[117] Meadows, however, clearly captures this postcrisis condition, characterized by the need for the resistance movement to consolidate its gains and authority, structuralizing its role in stability through cathartic or official means.[118] Hopper further clarifies this phase in his proposed "institutional stage of legalization and societal organization," during which "the out group must finally be able to legalize or organize their power" to institutionalize as a permanent organization "that is acceptable to the current mores." Additionally, the mechanisms at play within the movement have developed and compounded to become "well-nigh innumerable."[119]

Maoist and derivative phasing constructs regard the institutional state of specifically violent resistance movements in the consolidation, transition, and "regaining lost territories" phases.[120] Likewise, the CIA guide considers this state only within the "resolution stage" of insurgencies, during which any prolonged standoff or institutionalized perpetuation of conflict or balance of power is categorized as part of the "open insurgency stage."[121] The institutional state is the most mature phase of resistance before resolution (either successful or otherwise), but it can persist almost indefinitely if resolution is not achieved.

The cases of Hizbollah in Lebanon, the Karen National Liberation Army (KNLA) in Burma, and the Revolutionary Armed Forces of Colombia (FARC) provide good examples of the institutionalization state. Each group was able to integrate into the fabric of society through institutions, services, economic stakes, and relationships (illicit in the case of the FARC) that established lasting roles and influence, although this integration was not without ongoing contentions and violence. Nonviolent resistance movements can also reach the institutional state. For instance, the Muslim Brotherhood in Egypt was driven underground by repeated crackdowns and bans, but it nevertheless persisted as an influential sociopolitical institution for more than half a century.

Hizbollah

Hizbollah in Lebanon is an example of a resistance group that remains in the institutional state. In July 1993, Hizbollah transitioned into the institutional state after a cease-fire ended the Seven-Day War against Israel. By this time, Hizbollah was perceived as a provisional authority and a legitimate representative of the Shia population in Lebanon. Additionally, Hizbollah operated as a political and paramilitary organization, participating in Lebanese elections as well as armed confrontation of Israel. For these reasons, Hizbollah transitioned into the institutional state as an equal opposition player with broadened appeal.

Hizbollah's organizational and strategic prowess deepened during this period as attacks against Israel and Israeli targets became more sophisticated, characterized by cyberattacks, rocket launches, terrorist activity, and war from 2006 to 2008. The group structuralized its role by controlling media outlets, including a satellite channel and several radio stations and newspapers, and signaled its consolidation of authority and gains in domestic support with significant electoral victories in 2009. Hizbollah has demonstrated its continued staying power through its recent involvement in the Syrian civil war, fighting with Assad against Sunni rebels, and in domestic political conflicts in 2011, 2013, and 2014. Hizbollah is internationally recognized as a political arm within Lebanese politics, with only the armed wing considered a terrorist organization; this recognition further signals the group's institutional status. Given Hizbollah's persistence and continued role as an equal opposition player, it has not reached a resolution state.

Karen National Liberation Army

The KNLA in Burma experienced the institutional state during two periods: 1962 to March 1988 and 1989 to the present. During the first period, the Karen resistance persisted through more than a decade in the crisis state as the Karen National Union (KNU). In 1975, the KNLA formed and took over as the armed branch of the KNU. During this period, the KNU and KNLA operated as a quasi-government along the Thai–Burmese border, providing social services, law enforcement, and provisional government. The resistance deepened its appeal and expressed its stay-ing power through a Karen resistance culture, including a national flag, a coat of arms, dress, an anthem, and a history curriculum. During this period, KNLA orga-nizational prowess peaked, as marked by the resistance movement's highest levels of troops and record profits from trade and taxation in controlled areas. The resis-tance returned to the crisis state after launching an uprising in March 1988 followed by a year of demonstrations and confrontation with the Burmese government.

Transition to the second period of the institutional state occurred when the KNLA refused a cease-fire and became the principal target of the Burmese Army in 1989. The resistance strengthened in this crisis state, primarily because of its reputation as being unwilling to negotiate with the government, and it deepened its stance as an equal opposition player through continued attacks on the government. This period was marked by cease-fires, concessions, and renewed confrontation, includ-ing a cease-fire in 2007 that led to withdrawals of Burmese forces from Karen-designated border areas and a ramped-up Burmese offensive in 2009. The KNLA (and the KNU) continued to receive significant external and domestic support, especially among the Karen population and diaspora. In October 2015, the KNLA signed a ceasefire agreement, known as the Nationwide Cease-fire Agreement, with the Burmese (Myanmar) government and seven other armed groups.[122] How-ever, the agreement did not require disarmament, and it did not answer how to determine the balance of power between the central government and the ethnic regions, such as Karen.[123] The agreement established federalism as the guiding principle for the intended next stage of negotiations in the peace process, known as the Panglong Peace Conference.[124] In January 2018, the KNLA announced it would not attend the Panglong Peace Conference meeting.[125] That meeting has been delayed repeatedly because of continuing ethnic tensions and their imped-ing of dialogues that lead to the conference.[126] That the KNLA has signed a cease-fire and participates as a party in an ongoing national peace conference process demonstrates that it continues as of this writing in the institutional state without a resolution.

Revolutionary Armed Forces of Colombia

The institutional state of the FARC started in February 1991 and extends to the present. FARC was strengthened in the crisis state by successful insurgent opera-tions that went largely unanswered by the Colombian government. Unchallenged by the government, FARC established formal training centers, a strategic leader-ship arm, and a military academy during this period. The resistance also deepened its strategic prowess and consolidated its authority by developing relationships with narcotraffickers through the early 1990s. These relationships provided the resis-tance with critical resources, most notably funding and regional bases of opera-tion. In 1998, FARC carried out attacks on the Colombian Army and antinarcotic forces, but again, the government took a passive stance in response. President Pas-trana's administration (1998–2002) ceded territory to FARC by declaring a demili-tarized zone known as the Despeje, allowing FARC to effectively control large areas in southern Colombia. Within this region, FARC served as a pseudo-government, providing social, health, and educational services to civilians and collecting taxes from farmers and drug traffickers.

President Juan Manuel Santos pursued peace negotiations with FARC that led to a deal signed by both parties and approved by the Colombian national legislature. The road to that success, however, was tumultuous. In December 2015, FARC announced a ceasefire, and in June 2016, negotiators from FARC and the Colombian government agreed on a disarmament roadmap, only to be threatened by clashes between the two in July 2016. August 2016 brought a new bilateral ceasefire. By the following month, both parties signed off on a peace accord in Cartagena. In October 2016, however, the Colombian government put the new peace deal to the public in a referendum, and the public voted against it by a slim majority. Negotiators met with critics of the deal in October and November to revise the terms. By November 12th, the negotiators from each side signed a revised deal in Havana, and on November 24, 2016, President Santos and FARC leader Londoño signed the new deal in Bogota. Instead of submitting the agreement to the Colombian people for approval, President Santos sent the agreement to the Congress for its approval. Congress approved the revised peace deal six days later. The implementation of the agreement has not been without its problems. In 2017, FARC forces voluntarily disarmed by turning in all weapons to a United Nations mission and proceeded to "territorial spaces for training and reincorporation."[127] FARC as an entity converted into a political party called the Common Alternative Revolutionary Force, which under the agreement is guaranteed five seats in the first two electoral cycles in 2018.[128] It won only 0.3 percent of the vote in the first cycle.[129] Though appearing successful, the process has also faced setbacks: Congress has not passed laws needed to fulfill promises made in the peace deal; former FARC members face difficulty finding employment, training for employment, and land; and it has been reported that some former FARC members in the demobilization zones have returned out of frustration to their previous life and formed dissident groups.[130] Becoming a political party secured FARC in the institutional state, but the tenuous nature of the current peace and the leaching of former members from reintegration camps back into dissident groups demonstrates that FARC has not fully reached a resolution state as of this writing.

Abeyance: Demobilization to Incipience

Although not a state of resolution, abeyance (sometimes referred to as dormancy[131]) is when the resistance group or movement practices "little or no mobilization," instead reverting to a coalescent or incipient state of "inward . . . focus on identity or values,"[132] while largely avoiding decisive confrontations and reducing recruitment efforts. In their article "Missed Opportunities: Social Movement Abeyance and Public Policy," Traci M. Sawyers and David S. Meyer contend:

During abeyance, movements sustain themselves but
are less visible in interaction with authorities. At the
same time, values, identity, and political vision can
be sustained through internal structures that permit
organizations to maintain a small, committed core of
activists and focus on internally oriented activities.[133]

Verta Taylor similarly theorizes that a movement's abeyance provides
"a measure of continuity for challenging groups," allowing them to
"succeed in building a support base and . . . influence" despite being
"confronted with a nonreceptive political and social environment."[134] A
resistance movement, despite falling back into a less mobilized coalescent or incipient state through abeyance, can reemerge and remobilize
after reinforcing its group identity and developing a larger support base.

Resolution States

Decline

The resolution state includes the decline of the resistance, represented by many specified types and ways in which such movements end.
Any resistance can enter resolution from any other state, although some
resolutions decline from particular states.[135] Referred to as "decline"
in social movement theory,[136] the title "resolution" is derived from the
CIA's guide, which outlines several outcomes in the "resolution stage."[137]
The Maoist and related derivative phasing constructs do not specify a
diversity of resolution states, instead assuming insurgent or guerrilla
success and consolidation of power.[138] Below are the variety of resolution states along with examples for a few states derived from the ARIS
case study volumes.

Radicalization

According to Tarrow, radicalization is "a shift in ideological commitments toward the extremes and/or the adoption of more disruptive
and violent forms of contention."[139] While radicalization is not a state
of decline in and of itself, Tarrow contends that it can be a "mechanism for demobilization" that is often simultaneous with, and reacting to, the decline of another rival wing of the same movement via
institutionalization.[140] Tarrow contests that radicalization and institutionalization often occur simultaneously: as one wing moderates, the

other radicalizes further toward nonnegotiable positions and its tactics become more escalatory, confrontational, and possibly violent.

Institutionalization

According to Tarrow, institutionalization is, in contrast to radicalization, "a movement away from extreme ideologies and/or the adoption of more conventional and less disruptive forms of contention." The process of institutionalization is characterized by a group seeking "accommodations with elites and electoral advantage" by moderating its tactics and goals.[141] Hopper referred to this finalization of resistance as "the institutional stage of legalization and societal organization," during which the group transforms itself into a permanent organization "that is acceptable to the current mores."[142] As noted earlier, this institutionalization can often occur simultaneously with radicalization among another wing of the group. While both lead to a decline in the movement, institutionalization may be seen as a success of the movement, at least partially. Depending on the perceived extent of this success, the resistance movement may lose its raison d'être.

Palestine Liberation Organization

The Palestine Liberation Organization (PLO) offers an example of resolution by institutionalization. The PLO entered into the resolution state through institutionalization in January 2006. Resolution through institutionalization is characterized by the resistance seeking accommodations with power-holding elites and adopting more conventional forms of contention. The PLO made a slow transition from the institutional state to resolution through institutionalization. Throughout the institutional state, the PLO moderated its tactics and shifted from armed resistance to diplomacy and bureaucratization, most notably through its recognition of Israeli statehood, participation in the 1993 Oslo Accords, and creation of the Palestinian Authority. This moderation of tactics led to a decline in popular support over time and allowed the more radical Hamas to gain footing among the public. The shift in popular support to the more radical Hamas was evidenced by its electoral control of the Palestinian Legislative Council after 2006 elections and violent takeover of the Gaza Strip in 2007. The victory of Hamas in the January 2006 elections signaled the PLO's transition into the resolution institutionalization state. The PLO continues to act as a representative for the Palestinian movement, especially among international audiences. In 2010, the PLO agreed to US-mediated talks with Israel, seeking to gain autonomy for Palestine. As of this writing, the PLO is in operation but largely through diplomatic, institutionalized channels rather than armed resistance.

Repression

Both Miller and Tarrow highlight repression as a resolution state. Miller establishes that "repression occurs when agents of social control use force to prevent movement organizations from functioning or prevent people from joining the movement organizations." The tactics of repression are numerous, including indictment, infiltration, physical attacks, harassment, threats to job and school access, the spread of false information, and "anything else that makes it more difficult for the movement to put its views before relevant audiences."[143] Tarrow further discusses the likely outcomes of repressive actions. While repression could lead to resolution through repression, it can also "push radicals into more sectarian forms of organization and more violent forms of action, and can push moderates into the arms of conservatives."[144] Thus, repression becomes a resolution state when the government uses it in such a way that it effectively halts the resistance.

Facilitation

According to Tarrow, facilitation is when the government or its vested interests bring about the decline of a resistance group or movement by satisfying "at least some of the claims of contenders." Opposite but related to repression, facilitation may be pursued to a limited degree and combined with measures of repression.[145] When the government facilitates some but not necessarily all of the resistance group's claims, this may have the effect of splitting the resistance movement. Facilitation may attract moderates to legitimate action or elites to satisfaction with the government response, while frustrating radicals who want more change. Such a split may weaken the resistance if it coincides with a decline in popular support. Tarrow states that governments often use facilitation with selective repression as an effective means to end a resistance movement.

Provisional Irish Republican Army

The PIRA transitioned into the resolution state on April 10, 1998, through facilitation. The facilitation resolution state is marked by a decline in the resistance movement after the government satisfies some of the resistance movement's claims or demands. PIRA declined when the 1998 Good Friday Agreement satisfied some of its demands. The agreement enacted policing reforms, mandated release of political prisoners, set up provisions for a popular vote on Northern Ireland's status, and established power-sharing institutions, in line with PIRA's demands. After the agreement, Sinn Féin, the political arm of PIRA, became one of largest parties in

Northern Ireland, remaining active to this day. The agreement also disarmed the PIRA, and in 2005, international observers announced the group's complete demobilization. There was popular support for the agreement, with 71 percent of voters in Northern Ireland and 94 percent in Ireland voting in its favor in 1999. Despite these achievements, the movement's primary goal of an independent and unified Ireland was not met, and the resistance entered the resolution state.

Success

As a state of decline, success is "a bit more complicated" according to Miller. Although one could imagine a resistance that sets particular goals, achieves them, and then subsides, it is more common for movements to be "forced into compromises that only sometimes are advantageous to the movement," and although they obtain "concessions from the dominant system, movement organizations often have to relinquish some portion of their claim to represent an independent radical opposition." This dynamic of absorption soon transforms what was once a resistance movement into an interest group, brought "into the structure of interests in the polity."[146] The shape of success, and the concessions required, can also reveal internal fractures within the resistance movement, likewise causing impotence and decline. Some members of the resistance movement may see success when certain goals are achieved, but others may perceive success only when the movement continues to grow. However, growth may also lead to the addition of new members who are less committed to the original resistance than are older members, and this may lead to factions that may weaken the movement overall.

Orange Revolution

The Orange Revolution is an example of a resistance movement that transitioned to the resolution state through success. Resolution through success does indicate some degree of fulfillment of resistance goals, but it also indicates the decline of the resistance in response to those successes. The Orange Revolution transitioned from the crisis state to resolution state through success on December 26, 2004, when Yushchenko, the resistance's candidate, won by a clear margin in a third election. After Yanukovich waged a prolonged legal battle, the Supreme Court upheld Yushchenko's electoral victory and he was sworn in as Ukraine's president on January 23, 2005, signaling the successful resolution of the resistance. The resistance further broke down and splintered after the decisive election because of the absence of a unifying enemy (Kuchma/Yanukovich), and ultimately Yanukovich was elected president in 2010, defeating former resistance leader Tymoshenko. The reemergence of the resistance is being debated in light of the 2014 uprisings in Kiev, the ousting of Yanukovich, and the ongoing conflict in eastern regions of the country.

Failure

A resistance movement's failure at the internal organizational level can threaten the movement as a whole, potentially leading to either dormancy or a more decisive resolution state if no other organizations take up the torch. The failure resolution state is particular to resistance organizations whose agency and conduct, rather than overpowering external conditions, are evident in the decline. According to Miller, "failure at the organizational level takes two major forms: factionalism and encapsulation."[147] First, "factionalism arises from the inability of the organization's members to agree over the best direction to take," leading to an organizationally fatal internal conflict. Second, "encapsulation occurs when the movement organization develops an ideology or structure that interferes with efforts to recruit members or raise demands," eventually causing a critical decline in mobilization and capabilities.[148]

Jackson et al. present four additional failure states specific to incipient movements, making a total of six. Although factionalism and encapsulation can conceivably cripple an organization in any state of resistance, some failures are particular to organizations in their incipiency. First, groups can fail by neglecting to establish "a preexisting network of communication linking those groups of citizens most likely to support the movement," effectively isolating themselves from growth or mobilization potential. Second, the "failure of an emergent leader to incorporate . . . [other] leaders into his organization" can stagger a burgeoning resistance group before it has matured. Third, the young movement may lack "a program to which a major section of the [participants] could give wholehearted support," stifling recruitment and internal commitment. Finally, failures that become "highly publicized" and "conspicuous," creating a fatally "weakened . . . public image" may result in the resistance movement's rapid failure (because the group is soundly discredited) or slow decline (because confidence in the group fails to recover).[149]

Co-Optation

According to Miller, "co-optation strategies are brought into play when individual movement leaders are offered rewards that advance them as individuals while ignoring the collective goals of the movement." The rewards or positions are meant "to identify the interests of [the resistance organization] with those of the dominant society."[150]

Movement organizations that "are highly dependent on centralized authority or on charismatic leadership" are particularly prone to this state of movement decline.[151] This process of co-optation, as proposed by Patrick G. Coy and Timothy Hedeen, includes the appropriation of means (language and technique), the assimilation of leadership and participants after limited inclusion and participation, the transformation of movement goals, and finally the regulation of enacted changes by the state or vested interests.[152]

Establishment with the Mainstream

John J. Macionis characterizes this state of decline by saying that a "movement may 'go mainstream' . . . [and] become an accepted part of the system—typically after realizing some of their goals—so that although they continue to flourish, they no longer challenge the status quo."[153] Although establishment with the mainstream is similar to institutionalization, when a movement enters this state, it is accepted as a voice within the dominant power structure, but it simultaneously avoids being co-opted by the dominant power.

Frente Farabundo Martí para la Liberación

The Frente Farabundo Martí para la Liberación Nacional (Farabundo Martí National Liberation Front, or FMLN) in El Salvador provides a useful example of resolution by establishment with the mainstream. The FMLN transitioned into the resolution state on January 16, 1992, through establishment with the mainstream after signing a peace accord with the Salvadoran government. Peace negotiations leading up to the 1992 accord would have been unlikely without both the growing influence of moderates within the FMLN who saw violence as unsustainable and unlikely to bring victory and the increasing exhaustion among the landed elite who suffered economically during the civil war.

By signing the accord, the FMLN accepted concessions from the government, most notably gaining recognition as a political party, allowing it to enter the mainstream. The accord addressed some of the FMLN's other critical demands by enacting land reforms to help the peasant class, creating an independent body (the United Nations [UN] Truth Commission for El Salvador) to investigate atrocities carried out during the war, establishing a civilian police force, and placing constitutional limits on the military's power. Lastly, the accord outlined the demobilization of both the FMLN and the Armed Forces of El Salvador, and demobilization was carried out under UN observation over eighteen months after the signing of the accord. Today, the FMLN operates as one of the largest parties in El Salvador.

Exhaustion

After a resistance movement has progressed to a more mature state, particularly in the face of an extended crisis state, the organization and other participants may experience gradual decline through "psychological exhaustion which undermines the emotional foundations of the revolution."[154] As Edwards contends, this slow deflation of zeal for resistance and exhaustion is likely to be characterized by the eventual success of moderate or established interests, progressing directly toward a return to normalcy.[155] Tarrow also outlined exhaustion as a state of decline, stating that "although street protests, demonstrations, and violence are exhilarating at first . . . [resistance movements] involve risk, personal costs, and, eventually, weariness and disillusionment." The consequential and unequal decline in participation resulting from this dynamic poses unique challenges to resistance leaders and can contribute to movement radicalization or institutionalization.[156]

Now that the literature has been considered and a synthesis of phasing constructs from that literature proposed, this report now turns to demonstrating how one can use this proposed construct to study resistance movements. It is hoped that future, more in-depth analysis using this construct will be able to shed significant light on the mechanisms and variables that drive resistance movements forward and backward. For the purposes of demonstration, a limited analysis was undertaken and is presented in the following section. The cases are from the ARIS *Casebook on Insurgency and Revolutionary Warfare*, volume I: 1927–1962 and volume II: 1962–2009. Each of the forty-six cases in those two volumes has been coded according to the construct proposed in this report, and the appendix includes brief explanations of their coding.

ANALYSIS OF ARIS CASE STUDIES AS A PROOF OF CONCEPT

Coding the progression of ARIS case studies through the proposed states allows for those cases to be sorted based on their trajectory through the states. Knowing the trajectory of cases through the proposed phases allows a researcher to systematically compare and contrast cases and conduct more targeted, in-depth analyses that can shed light on the mechanisms and variables that impact the growth of a resistance. Patterns that appear not only in the coding of the cases but also in their specific characteristics in each state could indicate mechanisms that enable a movement to progress from one phase to the next or to regress to an earlier state. A key to the coding follows:

- P = Preliminary state
- I = Incipient state
- C = Crisis state
- N = Institutional state
- R = Resolution state
- r = Radicalization
- i = Institutionalization
- p = Repression
- f = Facilitation
- s = Success
- l = Failure
- Six different manifestations of a failure state of resolution are possible and detailed in the text (factionalism, encapsulation, failure in preexisting network, failure to incorporate other leadership, lack of program for enthusiastic support, and highly publicized failures). Coding summaries should specify which form of failure applies to each relevant case.
- c = Co-optation
- m = Establishment with the mainstream
- e = Exhaustion
- > = Progression in Phasing
- ^ = Reversal in Phasing

Cases are grouped together according to their coding so that cases with the same trajectory are presented together. Comparing and contrasting the cases that share trajectories, a brief assessment focuses on the possible mechanisms and variables that impacted the trajectories of those resistances. Table 3 presents all the coded cases grouped according to their path through the proposed construct. The authors recognize that a selection bias exists because of the use of only the case studies available in the ARIS body of work. However, these case studies are meant to be illustrative and not exhaustive or comprehensive of all instances of resistance. It is hoped that future research will use and test this phasing construct more rigorously.

Table 3. Full list of coded case studies grouped according to their path through the proposed construct.

Case Study	Analysis
Sudan Coup (1958)	P > I > R(s)
Congolese Coup (1960)	P > I > C > R(s)
Iraqi Coup (1936)	P > I > C > R(s)
Egyptian Coup (1952)	P > I > C > R(s)
Iranian Coup (1953)	P > I > C > R(s)
Iraqi Coup (1958)	P > I > C > R(s)
Iranian Revolution (1979)	P > I > C > R(s)
Korean Revolution (1960)	P > I > C > R(s)
Czechoslovakian Coup (1948)	P > I > C > R(s)
KLA (1996–1999)	P > I > C > R(s)
Orange Revolution (2004–2005)	P > I > C > R(s)
Hungarian Revolution (1956)	P > I > C > R(p)
Revolution in Vietnam (1946–1954)	P > I > C > N > R(s)
Guatemalan Revolution (1944)	P > I > C > N > R(s)
Argentine Revolution (1943)	P > I > C > N > R(s)
Tunisian Revolution (1950–1954)	P > I > C > N > R(s)
Afghan Mujahidin (1979–1989)	P > I > C > N > R(s)
Chinese Communist Revolution (1927–1949)	P > I > C > N > R(s)
Spanish Revolution (1936)	P > I > C > N > R(s)
FARC (1966–present)	P > I > C > N > R(i)
FMLN (1979–1992)	P > I > C > N > R(m)
LTTE (1976–2009)	P > I > C > N > R(p)
Revolution in French Cameroun (1956–1960)	P > I > C > N > R(p/f)
PIRA (1969–2001)	P > I > C > N > R(f)
Chechen Revolution (1991–2002)	P > I > C > N > R(r)
Hizbul Mujahideen (1989–present)	P > I > C > N > R(r)
Movement for the Emancipation of the Niger Delta (MEND; 2005–2010)	P > I > C > N > R(c)
Venezuelan Revolution (1945)	P > I > C > N > R(l—factionalism)

Case Study	Analysis
Egyptian Islamic Jihad (EIJ; 1928-2001)	P > I > C > N > R(1—encapsulation)
New People's Army (NPA; 1969–present)	P > I > C > N ^ I(a)
Hizbollah (1982–2009)	P > I > C > N
Shining Path (1980–1992)	P > I > C > N ^ C > R(p)
Viet Cong (1954–1976)	P > I > C > N ^ C ^ I > R(s)
Al Qaeda (1988–2001)	P > I > C > N ^ C ^ I
Taliban (1994–2009)	P > I > C > N ^ C > N
KNLA (1949–present)	P > I > C > N ^ C > N
Solidarity (1976–1990)	P > I > C ^ I > C > R(s)
Revolution in Malaya (1948–1957)	P > I > C ^ I > C > R(e)
German Revolution (1933)	P > I > C ^ I > C > R(s)
Hutu–Tutsi Genocides (1994)	P > I > C ^ I > C > R(f)
Revolutionary United Front (RUF; 1991–2002)	P > I > C ^ I > C > N > R(f)
PLO (1964–present)	P > I > C ^ I > C > N > R(i)
Bolivian Revolution (1952)	P > I > C ^ I > C > N > R(e/s)
Indonesian Rebellion (1945–1949)	P > I > C ^ I > C > N > R(s)
Cuban Revolution (1953–1959)	P > I > C ^ I > C > N > R(s)
Algerian Revolution (1954–1962)	P > I > C ^ I > C > N > R(s)

The first stage of resistance consists of general unrest that lacks an organized movement. It marks the beginning of a movement by providing the motivation for the creation of a resistance organization. Transition from this state can consist of the creation of an organization, which would mark the transition to the incipient state, or the unrest could be resolved without the development of a resistance organization or a crisis. The resolution of discontent in society through normal political processes is a fairly common and unremarkable event. Cases that move from the preliminary state to resolution are not represented in the ARIS case studies. Similarly, none of the ARIS cases regress from incipiency to the preliminary state. Social movements that begin to organize only to have the organization unravel without developing into a crisis or achieving some sort of resolution would be considered unremarkable cases and are not represented in the ARIS studies. By focusing on resistance movements, ARIS presupposes that there is a movement and this assumption requires incipiency.

Table 4. Intraelite conspiracy—resolution without crisis.

Case Study	Analysis
Sudan Coup (1958)	P > I > R(s)

All of the ARIS cases begin with a transition from the preliminary state to incipiency to crisis except for the Sudan Coup of 1958. The Sudan Coup case represents an interesting outlier in the ARIS data as the only case that does not involve a crisis state. The resistance movement did not encounter any opposition from the government and went directly from planning to success. Transitions from the incipient stage to resolution may result from successful cooperation between the resistance movement and the existing authorities. In the case of Sudan in 1958, the coup was a conspiracy between the conservative military and the conservative traditional leaders who had previously backed the government. The government cooperated in the conspiracy, and it was able to maintain the same social structure while reforming the government and making small policy changes to prevent more radical revolution. Conspiracies of this sort obviate the need for a crisis phase between incipiency and resolution.

Table 5. Coups and popular revolutions—short crises with decisive resolutions.

Case Study	Analysis
Congolese Coup (1960)	P > I > C > R(s)
Iraqi Coup (1936)	P > I > C > R(s)
Egyptian Coup (1952)	P > I > C > R(s)
Iranian Coup (1953)	P > I > C > R(s)
Iraqi Coup (1958)	P > I > C > R(s)
Iranian Revolution (1979)	P > I > C > R(s)
Korean Revolution (1960)	P > I > C > R(s)
Czechoslovakian Coup (1948)	P > I > C > R(s)
KLA (1996–1999)	P > I > C > R(s)
Orange Revolution (2004–2005)	P > I > C > R(s)
Hungarian Revolution (1956)	P > I > C > R(p)

The cases listed above follow a similar trajectory, having short crises that are quickly followed by resolution. With the exception of the Hungarian Revolution, all resulted in success. These cases can be divided into two types: coups and popular revolutions. Any similarities in their progression through the phases can be attributed to the planning that these movements carry out in the incipient state. The following examination of the cases provides insight into how these cases progressed from incipience through crisis and into resolution.

The coups in Congo, Egypt, Iran (1953), and Iraq (both 1936 and 1958) were all organized by the militaries in those countries. Their planning processes were similar, with a small group of officers organizing to overthrow the government. The coup in Egypt is slightly different from the others in that it was carried out by relatively junior officers who had created a formal organization that had existed for a considerable amount of time. The other coups were planned and executed by senior officers in relatively short periods of time. The militaries of these countries were well equipped to plan and execute revolutionary action because they were armed and organized. The outcome was assured once there was agreement among the key figures within the military. The ease of organizing and executing a military coup, as opposed to a revolution organized out of society at large, can contribute to why coups are often quick and successful. Not all coups are successful, however, as will be seen in the example of the Spanish Civil War, which began as a coup, in the next section.

The Czechoslovakian Coup was a communist takeover of a socialist government. It was not a military coup, but it followed a trajectory similar to that of a military coup. The minister of the interior carried out the revolution by using police forces to extract concessions from the government. The country was already closely tied to the Soviet Union, but it was not completely under communist control. Conservative parties had been banned after World War II, and within the left-leaning government, the communist minister of the interior appointed communists to key positions in the security services. The minister of the interior then used the police to force the prime minister to accept a communist government, which proceeded to achieve the movement's goals by changing the constitution and aligning Czechoslovakia with the Soviet Union.

The revolutions in Korea, Ukraine, Hungary, and Iran (1979) resulted from popular protests without the military playing a central role. These revolutions resulted from dissatisfaction among the people manifesting in mass protests and mob violence that overwhelmed the government's ability to suppress the resistance. Revolutions of this type can continue only for a limited time without resulting in either a resolution or the development of a resistance organization that is capable of resisting government suppression. If the protesters were unable to achieve their goals during the crisis period, they would either fall back into incipiency or develop into an institutional state. If the latter, the movement would consolidate its gains and develop into an organization with long-term staying power. Examples of both of these cases are discussed below.

The Kosovo Liberation Army (KLA) differs somewhat from the other cases in that it was neither a coup nor a popular revolution, but instead a military resistance organization. The KLA organized over several years and then fought the Serbian government in a civil war. This case is similar to many of the cases that include institutionalization, except the relatively short duration of the Kosovo War prevented the institutionalization of this conflict.

Because most of the movements in the ARIS sample were successful, this may imply that decisive crises are more likely to succeed than longer conflicts are. However, this is not necessarily true. This document is limited to discussion of the ARIS case studies, and therefore the relationship between the transitions from crisis to success may be the result of the selection of these cases. The fact that crises resulting in failure are uncommon in this sample could be the result of them being uncommon in the real world or instead could be the result of ARIS case studies having been selected to represent notable cases. For example, failed military coups might be of less interest than successful ones and would thus be less likely to be chosen for study.

Table 6. Crisis to institutionalization to resolution.

Case Study	Analysis
Revolution in Vietnam (1946–1954)	P > I > C > N > R(s)
Guatemalan Revolution (1944)	P > I > C > N > R(s)
Argentine Revolution (1943)	P > I > C > N > R(s)
Tunisian Revolution (1950–1954)	P > I > C > N > R(s)
Afghan Mujahidin (1979–1989)	P > I > C > N > R(s)
Chinese Communist Revolution (1927–1949)	P > I > C > N > R(s)

Case Study	Analysis
Spanish Revolution (1936)	P > I > C > N > R(s)
FARC (1966–present)	P > I > C > N > R(i)
FMLN (1979–1992)	P > I > C > N > R(m)
LTTE (1976–2009)	P > I > C > N > R(p)
Revolution in French Cameroun (1956–1960)	P > I > C > N > R(p/f)
PIRA (1969–2001)	P > I > C > N > R(f)
Chechen Revolution (1991–2002)	P > I > C > N > R(r)
Hizbul Mujahideen (1989–present)	P > I > C > N > R(r)
MEND (2005–2010)	P > I > C > N > R(c)
Venezuelan Revolution (1945)	P > I > C > N > R(l—factionalism)
EIJ (1928–2001)	P > I > C > N > R(l—encapsulation)
NPA (1969–present)	P > I > C > N ^ I(a)
Hizbollah (1982–2009)	P > I > C > N

Whereas the previous cases quickly reached resolution from the crisis state, other cases move from crisis to institutionalization and then reach a resolution. In these cases, the crisis itself was not decisive, leading the resistance movements to deepen their organizational and strategic capability while continuing their struggle. Most of these cases are civil wars in which the resistance develops a military organization that engages in conflict with the government for an extended period. The resolutions that end these conflicts are more diverse than the shorter coups and revolutions that resolve directly from the crisis state. The complex situations that evolve over time during the institutional phase can result in a variety of resolutions, such as the movement's success, its institutionalization into the existing system, its joining the mainstream, its repression, the government's facilitation of its demands, its radicalization, co-optation of its leaders, its organizational failure through encapsulation, or its reversion into a state of abeyance. In all of these cases, the crises do not lead directly to resolution, and instead the resistance movement develops and progresses into the institutionalization phase.

Most of the movements that progress from crisis to institutionalization are armed insurrections against the government. These cases differ from those of movements that progress directly from crisis to resolution in that they generally involve neither large protests with the support of the population nor support from the military. The Afghan Mujahidin, Chinese Civil War, FARC, FMLN, LTTE, the resistance in French Cameroun, PIRA, the Chechen War, Hizbul Mujahideen, MEND, EIJ, NPA, and Hizbollah began with militant organizations fighting armed insurrections against the government. As these conflicts progressed from small insurrections into large-scale wars, organizations developed, and thus institutionalization occurred. Institutionalization therefore appears to be more likely to follow crisis when the resistance movement is attempting to seize power through military means. This is likely the result of decisions made in the incipient stage when the resistance determines how to organize. When the government actively suppresses dissent, resistance movements are more likely to develop into small clandestine organizations, relying on violent methods, whereas in less oppressive situations, a movement might attempt to gain popular support and mobilize supporters into a popular revolution. In other cases, a group's decision to adopt violent methods of revolution may be based on its lack of popular support in the population, such as when the organization is based on an ideology that is not popular enough to support large-scale protests. These factors force decisions early in the development of the resistance that carry through to later stages. For example, the decision to create a violent revolutionary movement frequently leads to the institutionalization of that movement, which often occurs years later after the conflict has progressed considerably.

Although most cases of institutionalization begin with insurgencies that develop into larger organizations, some follow other courses. The Spanish Civil War began as a coup that was only partially successful and then progressed into a full-scale war. The initial planning was for a quick coup that would take control of cities throughout the country. However, after initiation of the coup, the military split and the coup leaders failed to consolidate power. A war between the two factions of the army ensued. Clear lines of battle developed and the nationalist faction institutionalized, developing a government that ruled the portion of the country under its control. The institutionalization was marked by the nationalists engaging in international relations with foreign supporters and building links to the Catholic Church and the monarchy to legitimize their claims to power.

Guatemala presents an unusual case of a popular revolution that institutionalized. The revolution began with people protesting in the streets and crowds engaging in violence. After the government repressed the protests and the protesters increased violence, elements within the military attempted a coup, but they were unsuccessful. The conflict developed into institutionalization because the initial protests were unsuccessful, forcing the resistance to increase its level of violence and attempt to cooperate with the military to overthrow the government. The conflict was eventually resolved through elections, which the opposition was able to win. The Tunisian Revolution followed a path similar to that of Guatemala: it was organized by a popular pro-independence organization that had attempted to use peaceful means, which were unsuccessful, forcing it to turn to a campaign of violence against the French authorities. Both of these cases illustrate a popular political party that developed into a militant organization.

In the cases of Argentina and Venezuela, institutionalization was the result of internal struggles between the revolutionaries themselves rather than a struggle between the resistance and the government. In both cases, military coups were able to unseat the previous government. However, the coups were followed by periods of institutionalization in which internal struggles within the new governments ensued. The Argentine Revolution was a military coup that succeeded quickly, but power struggles within the military led the crisis event of the coup to develop into a state of institutionalization because the military leaders struggled internally for power. One of those military leaders, Juan Perón, eventually consolidated power, resulting in a successful resolution.

The Venezuelan Revolution of 1945, however, reached a resolution state of factionalism instead of success. Similar to the Argentinian experience, the Venezuelan Revolution combined a popular movement with a military coup. In this case, the Democratic Action (AD) party attempted for several years to gain power through legal means, but when it did not succeed, it worked with the military to organize a coup sustained by large protests from party supporters. This constituted the crisis state. When the AD party installed itself as the new government, it reached institutionalization. However, when it asserted control over the military, it reached a resolution state of factionalism when the military launched another coup in response.

Table 7. Organizational destruction without outright defeat.

Case Study	Analysis
Shining Path (1980–1992)	P > I > C > N ∧ C > R(p)
Viet Cong (1954–1976)	P > I > C > N ∧ C ∧ I > R(s)
Al Qaeda (1988–2001)	P > I > C > N ∧ C ∧ I
Taliban (1994–2009)	P > I > C > N ∧ C > N
KNLA (1949–present)	P > I > C > N ∧ C > N

The civil wars discussed above reached conclusions after institutionalizing. However, another path for an institutionalized resistance organization is a return to crisis. These cases represent institutionalized groups that faced crises that destroyed their institutional structure but did not result in complete failure. The Shining Path eventually succumbed to government repression, whereas the Viet Cong degraded considerably but was able to continue in a state of incipiency until it succeeded in taking over the government of South Vietnam. Al Qaeda, the Taliban, and the KNLA all faced major crises but still continue their struggles. These cases may represent a special type of civil war, presenting groups that are able to withstand substantial degradation through military defeat while continuing to operate. An examination of the crises that pushed these organizations out of the institutional phase, and why they were able to survive these crises, will provide insight into the qualities of the crises, and of the organizations, that allowed them to survive.

Shining Path and the Viet Cong were both large, successful military organizations that experienced crises due to military losses that threatened their existence. Shining Path has been all but defeated and is no longer a serious threat to the government of Peru, whereas the Viet Cong was able to succeed only in overthrowing the government of South Vietnam because of the victory of North Vietnamese Army.

Shining Path developed into a large, institutionalized resistance movement, controlling a large portion of Peru. However, starting in 1989, Shining Path increasingly used terrorist methods to attack urban targets, damaging its support among the population. Government efforts to combat Shining Path increased, and through a three-year crisis period the organization degraded. The leader of Shining Path, Abimael Guzman, was captured, and although Shining Path still exists, it has been reduced so significantly that the conflict can be considered resolved. This case shows that an institutionalized organization can face a crisis from which it cannot recover. This crisis was brought on in part by the success of Shining Path. Its victories in rural areas led it to expand operation to urban areas, which in turn shifted public opinion against the group and provoked increased government response.

The Viet Cong was able to develop into an institutionalized force but faced a crisis during the Tet Offensive that reduced it to an incipient state. The movement's military losses amounted to about half of its total fighting force, causing it to cease being a military force for the remainder of the war. The success of the North Vietnamese Army allowed the Viet Cong to succeed in achieving its objective even though the group was no longer functioning as a military force.

Al Qaeda has followed a trajectory similar to that of Shining Path and the Viet Cong, with a crisis reducing it from institutionalization to incipiency, but it still continues to function. Al Qaeda became an institutionalized terrorist organization in the early 1990s and continued to function until the US invasion of Afghanistan in 2001 forced it into a crisis state once again. US military action was able to degrade Al Qaeda's ability so that it is no longer a functional organization, although it still provides an influential philosophy and ideology. Al Qaeda is now in an incipient state, limited to organizing and facilitating other organizations rather than carrying out its own operations.

Unlike the movements described above, the Taliban and the KNLA faced crises but were able to recover from their respective crises to reinstitutionalize, and both continue to exist as institutional resistance movements. After the US invasion of Afghanistan, the Taliban was forced from power and its leaders fled to Pakistan to reorganize. However, the movement was able to reconstitute, and it returned to the institutional phase again by 2003. After consolidating power in the border region of Pakistan, the Taliban has been able to resume operations in Afghanistan, particularly since the withdrawal of US forces. The Taliban's ability to recover after the US invasion could be explained by its widespread support among the Pashtun people and its ability to seek safe haven in Pakistan due to the existence of Pashtun populations on both sides of the border. The Taliban is composed almost entirely of Pashtuns, and the support from ethnic Pashtuns in Pakistan provides the group with a level of resilience that other organizations may lack.

The KNLA has been fighting the Burmese government for decades and is highly institutionalized. In 1988, the KNLA, along with other Burmese resistance groups, supported democratic protests that were brutally suppressed by the government. These protests were followed by a military coup that prompted most of the ethnonationalist groups in Burma to agree to cease-fires with the government. The KNLA refused a cease-fire and became the principal target of the Burmese military. After this crisis, the KNLA was able to recover and continues to fight the government.

The cases of institutionalization regressing to crisis provide two possible trajectories. Either the crisis signals the beginning of a decline in the organization with a return to incipiency or resolution through repression or the crisis simply represents a setback from which the organization is capable of recovering. The nature of the crisis and the environment in which the organization operates appear to be the most likely determinants of whether an organization will be able to recover from a postinstitutional stage crisis.

Those organizations just discussed went through crisis phases that did not result in their institutionalization or resolution but instead their return to incipience. Crises present the potential for resolution of the conflict or institutionalization of the conflict. A group's transition from crisis to incipiency represents its failure to secure victory but also its success at maintaining the revolutionary organization. These resistance movement organizations persevered through crisis to maintain and reorganize their movements. After regressing to incipiency, they all returned to the crisis state again, ultimately attaining their goals through outright success as well as facilitation and institutionalization. An investigation of the transition from crisis to incipience, including the transformations occurring in incipience that allowed for success in the second crisis, will demonstrate why this trajectory so often results in successful outcomes.

Table 8. Failed crises followed by repeat attempts.

Case Study	Analysis
Revolution in Malaya (1948–1957)	P > I > C ^ I > C > R(e)
German Revolution (1933)	P > I > C ^ I > C > R(s)
Hutu–Tutsi Genocides (1994)	P > I > C ^ I > C > R(f)
Solidarity (1976–1990)	P > I > C ^ I > C > R(s)
RUF (1991–2002)	P > I > C ^ I > C > N > R(f)
PLO (1964–present)	P > I > C ^ I > C > N > R(i)
Bolivian Revolution (1952)	P > I > C ^ I > C > N > R(e/s)
Indonesian Rebellion (1945–1949)	P > I > C ^ I > C > N > R(s)
Cuban Revolution (1953–1959)	P > I > C ^ I > C > N > R(s)
Algerian Revolution (1954–1962)	P > I > C ^ I > C > N > R(s)

The German Revolution that resulted in Hitler coming to power was preceded by an earlier attempt, the Beer Hall Putsch, during which the Nazi Party attempted to seize control of the Bavarian State government by force. This early attempt at armed insurgency resulted in jail terms for most of the party's leaders and prompted them to reconsider their strategy. While in prison, Hitler formulated a plan to take power through legal means. This strategy worked well, and the Nazi party proved highly capable of gaining popular support. The second crisis did not start until Hitler was already chancellor and in a much better position to ensure his success in securing dictatorial power.

Similar cases of organizations attempting revolutions, failing, and then constructing improved plans for future attempts can be seen in Indonesia, Bolivia, Cuba, Algeria, and the RUF in Sierra Leone. In all of these cases, the organization initiated a crisis in an attempt to overthrow the government, failed, and was forced to retreat and reorganize. The return to incipience afforded the groups an attempt to regroup and reorganize. These organizations were then able to wait until opportunities presented themselves when they were then able to act, and in all of these cases the groups were successful. In Indonesia and Algeria, the movements were anticolonial in nature and the revolutionaries lived under the colonial regime while organizing, whereas in Cuba and Sierra Leone the resistance groups had to take shelter in foreign safe havens, in Mexico and Liberia, respectively. In all four cases, the experience of the first failed attempt at revolution allowed for stronger attempts in the second crisis period.

In Malaya, the first crisis the Malayan Communist Party (MCP) faced was Japan's occupation. The British worked with the party to expel the Japanese, and the MCP did not completely demobilize after the war. Although the group returned to incipiency, the first crisis had improved it. The experience of fighting the Japanese prepared the MCP to fight for independence, and it took the British nine years to wear down the resistance before they could transfer control to an independent noncommunist government. This is the only case of a successful conflict with an initial opponent, in this case the Japanese, preparing an organization to fight another enemy in a later episode.

In Rwanda, the Hutu–Tutsi genocide was preceded by an earlier crisis during which the Tutsi government's massacre of Hutus in Burundi led Tutsis in Rwanda to begin attacks on their Hutu government. This violence died down, returning the crisis to incipience. However, the seeds of resentment were planted and later contributed to the genocide in which the Hutu government massacred Tutsis. In this case, the movement from crisis to incipiency without resolution of the underlying conflict laid the groundwork for the renewal of a much more devastating crisis.

The PLO advanced to crisis when it began military attacks on Israel in 1965, but it returned to incipiency after the defeat of the 1967 war. The PLO reorganized under Yasser Arafat and returned to crisis in 1968 when it resumed attacks on Israeli targets. The return to incipience in this case was due to an interstate war that disrupted operations and forced the group to reorganize and reevaluate its strategy, moving the PLO away from conventional tactics and toward terrorism. The Six-Day War was unable to destroy the PLO and instead only disrupted its operations for a short time.

Solidarity is similar to those cases described above, but the nonviolent nature of this movement resulted in a less violent repression by the government. Solidarity entered a crisis state during which it confronted the government through strikes and protests, and after being repressed the organization spent several years waiting for another opportunity to act. When the next protest arose, protesters used largely the same techniques the first wave of protesters had used, and Solidarity met with success because the political climate was more receptive to its demands.

The cases of crisis returning to incipience were largely successful and progressed to a second crisis, leading to successful resolution, because the experiences of the first crisis prepared the organizations to succeed in the second one. Although the first crisis could be seen as a short-term failure, in that these groups entered a crisis with the government that did not allow them to achieve their goals, their ability to survive the crisis allowed for reorganization that contributed to later success. Taking a historical look at the trajectory of these movements through the phases of conflict shows that what appears as a failure at one point in time may actually serve to strengthen a movement in the end.

CONCLUSION

Resistance movements originate, grow, mature, escalate, and decline, changing in both shape and character during their progression. Scholars and analysts of resistance have long acknowledged the need to examine movements not only according to their unique features and behaviors but also according to their developmental phases. By conceptualizing and understanding a group's phases of development, unconventional warfare practitioners can better engage with a resistance movement.

This study has sought to contribute to furthering that understanding by synthesizing existing thought on the subject and by creating a construct that practitioners and scholars can use to further study the subject. The construct features five states: preliminary, incipient, crisis, institutional, and resolution. The most defining feature of the preliminary state is the manifestation of unorganized and unattributed unrest—unorganized because actors are unconnected, and unattributed because the unrest lacks a common narrative about the source of the problem. For the incipient state, alternatively considered coalescence, the defining feature is the development of intentional organization and a common narrative. The movement has begun to take shape, with leaders and overt, strategic patterns of action as opposed to featuring leaderless short-term actions. A movement in crisis state features the essential characteristic of an escalated confrontation with opponents (most often the state authority) that marks a decisive moment when the movement demonstrates itself to be a real and clear threat to the opponent's interests, such that the opponent must respond. In the institutional state, the resistance movement has survived the confrontation with its opponent and has established a role in society. Finally, a movement can reach a resolution state from any previous state through a variety of avenues, from falling dormant to radicalizing to exhausting its resources to being co-opted to succeeding and several more. This study graphically presents these states as linear, but properly understood the construct allows for the flexibility of resistance movements to progress in a nonlinear fashion. The appendix that follows uses existing ARIS case studies to demonstrate this feature, as well as how practitioners and scholars can use this construct to systematically study the concept of phases more deeply.

The ARIS body of work aims to further the science of resistance. The developmental framework (or "phasing construct") proposed in this work, coupled with the vast body of work under the ARIS umbrella, seeks to lay the foundation for a science of resistance that will not only support the needs of the Special Forces community but will also enrich the broader communities of academics and policy makers. It is our hope that others will use this construct to undertake extensive analysis that will shed significant light on the mechanisms and variables that drive resistance movements forward and backward.

NOTES

[1] Jonathon B. Cosgrove and Erin N. Hahn, *Conceptual Typology of Resistance* (Fort Bragg, NC: United States Army Special Operations Command, forthcoming).

[2] David J. Danelo, "Exploring the Phases of Contemporary Resistance," in *Special Topics in Irregular Warfare: Understanding Resistance*, ed. Erin Hahn (Fort Bragg, NC: United States Army Special Operations Command, forthcoming).

[3] Mao Tse-tung, *On Guerrilla Warfare* (Urbana: University of Illinois Press, 2000).

[4] Field Manual 3-24 (FM 3-24), *Counterinsurgency* (Washington, DC: Headquarters, Department of the Army, 2006).

[5] US Army Doctrine and Training Publication (ATP) 3-05, *Unconventional Warfare* (Washington, DC: Headquarters, Department of the Army, September 6, 2013).

[6] David Galula and John A. Nagl, *Counterinsurgency Warfare: Theory and Practice* (Westport, CT: Greenwood Publishing Group, 2006), 40.

[7] Ibid., 30.

[8] Andrew R. Molnar, *Human Factors Considerations of Undergrounds in Insurgencies* (Washington, DC: Special Operations Research Office, 1966), 2–3.

[9] Douglas McAdam, Sidney Tarrow, and Charles Tilly, *Dynamics of Contention* (New York: Columbia University, 2001), 24.

10 Maegen Gandy, "The Politics of Insurgency," (PhD diss., University of Maryland, 2015).

11 Erin N. Hahn and W. Sam Lauber, *Legal Implications of the Status of Persons in Resistance* (Fort Bragg, NC: United States Army Special Operations Command, 2014).

12 *Juan Carlos Abella v. Argentina*, Case 11.137, Report Number 55/97.

13 Hahn and Lauber, *Legal Implications*, 31.

14 Ibid., 40.

15 Ibid., 42.

16 *Prosecutor v. Tadic*, Case No. IT-94-1-AR72, Decision on Defence motion for Interlocutory Appeal on Jurisdiction, para. 70 (Int'l Crim. Trib. for the Former Yugoslavia October 2, 1995).

17 Anthony Cullen, *The Concept of Non-International Armed Conflict in International Humanitarian Law* (Cambridge: University of Cambridge, 2012), 10.

18 *Prosecutor v. Tadic*, para. 96.

19 Waldemar A. Solf, "The Status of Combatants in Non-International Armed Conflicts under Domestic Law and Transnational Practice," *American University Law Review* 33 (1983–1984): 59.

20 Robert D. Powers, "Insurgency and the Law of Nations," *JAG Journal* 16 (1962): 55–56.

21 *Prosecutor v. Tadic*, ¶ 70; see also Protocol Additional to the Geneva Conventions of 1949, and relating to the Protections of Victims of Non-International Armed Conflicts (Protocol II) art. 1(1), December 12, 1977, 1125 U.N.T.S. 609.

22 International Committee of the Red Cross, *Commentary, Convention Relative to the Treatment of Prisoners of War*, ed. Jean Pictet (Geneva: ICRC, 1960), 43–44.

23 Rosalyn Higgins, "International Law and Civil Conflict," in *The International Regulation of Civil Wars*, ed. Evan Luard (New York: New York University Press, 1972), 170–171; and Hersch Lauterpacht, *Recognition in International Law* (Cambridge: Cambridge University Press, 1948), 176.

24 Lothar Kotzsch, *The Concept of War in Contemporary History and International Law* (Geneva: Libraire E. Droz, 1956), 234.

25 Ibid.

26 R. T. Naylor, "The Insurgent Economy: Black Market Operations of Guerilla Organizations," *Crime, Law, and Social Change* 20, no. 1 (1993): 13–51.

27 Eli Berman, Jacob N. Shapiro, and Joseph H. Felter, "Can Hearts and Minds Be Bought? The Economics of Counterinsurgency in Iraq," *Journal of Political Economy* 119, no. 4 (2011): 766–819.

28 Joel M. Guttman and Rafael Reuveny, "On Revolt and Endogenous Economic Policy in Autocratic Regimes," *Public Choice* 159, no. 1 (2014): 27–52.

29 Lyford P. Edwards, *The Natural History of Revolution* (Chicago: University of Chicago Press, 1927), 23.

30 Ibid., 25.

31 Ibid., 38.

32 Bob Jessop, "Reviewed Work: *The Natural History of Revolution* by Lyford P. Edwards," *Sociology* 6, no. 1 (1972): 130.

33 Edwards, *Natural History of Revolution*, 98.

34 See a review in Torbjørn L. Knutsen and Jennifer L. Bailey, "Review Essay: Over the Hill? *The Anatomy of Revolution* at Fifty," *Journal of Peace Research* 26, no. 4 (1989): 421–431.

35 Crane Brinton, *The Anatomy of Revolution*, rev. ed. (New York: Vintage Books, 1965).

36 Ibid.

37 Ibid.

38 Ibid.

39 Ibid.

40 Paul Meadows, "Sequence in Revolution," *American Sociological Review* 6, no. 5 (1941): 707–709.

41 Rex D. Hopper, "The Revolutionary Process: A Frame of Reference for the Study of Revolutionary Movements," *Social Forces* 28, no. 3 (1950): 271–272.

42 Jonathan Christiansen, *Social Movements & Collective Behavior: Four Stages of Social Movements*, Research Starters Academic Topic Overviews (Ipswich, MA: EBSCO Publishing, 2009).

43 Hopper, "The Revolutionary Process," 271–272.

44 Ibid., 272–275.

45 Ibid., 275–277.

46 Ibid., 277–279.

47 Frederick D. Miller, "The End of SDS and the Emergence of Weatherman: Demise through Success" in *Waves of Protest: Social Movements since the Sixties*, ed. Jo Freeman and Victoria Johnson (Lanham, MD: Rowman & Littlefield Publishers, 1999), 303.

48 Ibid., 304.

49 Ibid.

50 Ibid.

51 Ibid., 304–305.

52 Ibid., 305.

53 Ibid., 306–307.

54 Ibid., 307–308.

55 Sidney G. Tarrow, *Power in Movement: Social Movements and Contentious Politics*, rev. and updated third ed. (Cambridge, NY: Cambridge University Press, 2011), 185.

56 Ibid., 187.

57 Ibid., 189.

58 Ibid., 190.

59 Ibid., 192.

60 Ibid., 197–198.

61 US Central Intelligence Agency, *Guide to the Analysis of Insurgency 2012* (Washington, DC: US Government, 2012).

62 Ibid., 5–9.

63 Ibid., 10–12.

64 Ibid., 13–16.

65 Ibid., 17–21.

66 Maurice Jackson et al., "The Failure of an Incipient Social Movement," *The Pacific Sociological Review* 3, no. 1 (1960): 40.

[67] Patrick G. Coy and Timothy Hedeen, "A Stage Model of Social Movement Co-Optation: Community Mediation in the United States," *The Sociological Quarterly* 46, no. 3 (2005): 405.

[68] Ibid., 410.

[69] Ibid., 411.

[70] Ibid., 413–420.

[71] Ibid., 420–423.

[72] Ibid., 423–426.

[73] Ralph H. Turner, "New Theoretical Frameworks," *The Sociological Quarterly* 5, no. 2 (1964): 122–132.

[74] Ibid., 126.

[75] Ibid., 127.

[76] Ibid., 129, discussing Gustave LeBon, *The Crowd: A Study of the Popular Mind* (London: Ernest Benn, 1896); Sigmund Freud, *Group Psychology and the Analysis of the Ego* (London: Hogarth Press, 1922); and Wilfred Trotter, *Instincts of the Herd in Peace and War: 1916–1919* (London: Oxford University Press, 1919).

[77] Michael Woods et al., "'The Country(side) Is Angry': Emotion and Explanation in Protest Mobilization," *Social & Cultural Geography* 13, no. 6 (2012): 567–585.

[78] Ibid., 571.

[79] FM 3-24, *Insurgencies and Countering Insurgencies*, 4–8.

[80] Christiansen, *Four Stages of Social Movements*.

[81] Edwards, *Natural History of Revolution*, 23–25.

[82] Brinton, *Anatomy of Revolution*.

[83] Meadows, "Sequence in Revolution."

[84] Hopper, "The Revolutionary Process," 271–272.

[85] James Chowning Davies, "The J-Curve and Power Struggle Theories of Collective Violence," *American Sociological Review* 39, no. 4 (1974): 607–610; and James C. Davies, "Toward a Theory of Revolution," *American Sociological Review* 27, no. 1 (1962): 5–19.

[86] T. Gurr, *Why Men Rebel* (Princeton, NJ: Princeton University Press, 1970).

[87] Davies, "Toward a Theory of Revolution."

[88] Thomas F. Pettigrew, "Samuel Stouffer and Relative Deprivation," *Social Psychology Quarterly* 78, no. 1 (2015): 7–24.

[89] R. Williams, "Relative Deprivation," In *The Idea of Social Structure: Papers in Honor of Robert K. Merton*, edited by L. Coser (New York: Harcourt, Brace Jovanovich, 1975).

[90] Pettigrew, "Samuel Stouffer and Relative Deprivation."

[91] Danelo, "Exploring the Phases of Contemporary Resistance," 11–13.

[92] US Central Intelligence Agency, *Guide to the Analysis of Insurgency*, 5–9.

[93] Hopper, "The Revolutionary Process," 272–275; Christiansen, *Four Stages of Social Movements*, 3; and Brinton, *Anatomy of Revolution*.

[94] Christiansen, *Four Stages of Social Movements*.

[95] Jackson et al., "The Failure of an Incipient Social Movement," 35–40.

[96] Hopper, "The Revolutionary Process," 272–275.

[97] Ibid., 273; and Christiansen, *Four Stages of Social Movements* 3.

98 Christiansen, *Four Stages of Social Movements*, 3.

99 Edwards, *Natural History of Revolution*, 38; and Jessop, "Reviewed Work: *The Natural History of Revolution*," 130.

100 Brinton, *Anatomy of Revolution*.

101 Meadows, "Sequence in Revolution."

102 Danelo, "Exploring the Phases of Contemporary Resistance," 11–13.

103 US Central Intelligence Agency, *Guide to the Analysis of Insurgency*, 10–11.

104 Ibid., 10.

105 Edwards, *Natural History of Revolution*; Brinton, *Anatomy of Revolution*; and Meadows, "Sequence in Revolution."

106 Hopper, "The Revolutionary Process"; and Christiansen, *Four Stages of Social Movements*.

107 Edwards, *Natural History of Revolution*, 98.

108 Brinton, *Anatomy of Revolution*.

109 Meadows, "Sequence in Revolution."

110 Hopper, "The Revolutionary Process," 275–277.

111 Ibid., 275.

112 Christiansen, *Four Stages of Social Movements*.

113 Danelo, "Exploring the Phases of Contemporary Resistance," 11–13.

114 US Central Intelligence Agency, *Guide to the Analysis of Insurgency*, 13.

115 Erica Chenoweth and Maria J. Stephan, "Drop Your Weapons," Foreign Affairs 93, no. 4 (July 2014): 94–106.

116 Christiansen, *Four Stages of Social Movements*.

117 Edwards, *Natural History of Revolution*; and Brinton, *Anatomy of Revolution*.

118 Meadows, "Sequence in Revolution."

119 Hopper, "The Revolutionary Process," 275–277.

120 Danelo, "Exploring the Phases of Contemporary Resistance," 11–13.

121 US Central Intelligence Agency, *Guide to the Analysis of Insurgency*, 13–17.

122 Wai Moe and Thomas Fuller, "Myanmar and 8 Ethnic Groups Sign Cease-Fire, but Doubts Remain," *New York Times*, October 15, 2015, https://www.nytimes.com/2015/10/16/world/asia/myanmar-ceasefire-armed-ethnic-groups.html.

123 Ibid.

124 Ibid.

125 Chit Min Tun, "KNLA Says It Won't Attend Third Sessions of Panglong Peace Conference," *Irrawaddy*, January 9, 2018, https://www.irrawaddy.com/news/burma/knla-says-wont-attend-third-session-panglong-peace-conference.html.

126 Nyein Nyein, "Third Session of Panglong Peace Conference Pushed Back to May," *Irrawaddy*, March 1, 2018, https://www.irrawaddy.com/news/burma/third-session-panglong-peace-conference-pushed-back-may.html.

127 Adam Isacson, "Colombia's Imperiled Transition," *New York Times*, April 5, 2018, https://www.nytimes.com/2018/04/05/opinion/colombia-farc-transition.html; "Colombia's FARC Officially Ceases to Be an Armed Group," *BBC News*, June 27, 2017, http://www.bbc.com/news/world-latin-america-40417207.

[128] Jim Wyss, "Colombia Signs New Peace Pact with FARC Guerillas," *Miami Herald*, November 24, 2016, http://www.miamiherald.com/news/nation-world/world/americas/colombia/article116872338.html.

[129] Isacson, "Colombia's Imperiled Transition."

[130] Ibid.

[131] Stacy Keogh, "The Survival of Religious Peace Movements: When Mobilization Increases as Political Opportunity Decreases," *Social Compass* 60, no. 4 (2013): 561–578.

[132] Christiansen, *Four Stages of Social Movements*, 6.

[133] Traci M. Sawyers and David S. Meyer, "Missed Opportunities: Social Movement Abeyance and Public Policy," *Social Problems* 46, no. 2 (1999): 188.

[134] Verta Taylor, "Social Movement Continuity: The Women's Movement in Abeyance," *American Sociological Review* 54, no. 5 (1989): 762.

[135] Jackson et al., "The Failure of an Incipient Social Movement."

[136] Christiansen, *Four Stages of Social Movements*, 3–4; Miller, "The End of SDS."

[137] US Central Intelligence Agency, *Guide to the Analysis of Insurgency*, 17–21.

[138] Danelo, "Exploring the Phases of Contemporary Resistance," 11–13.

[139] Tarrow, *Power in Movement*, 207.

[140] Ibid., 190, 207–208.

[141] Ibid., 207–208.

[142] Hopper, "The Revolutionary Process," 277–279.

[143] Miller, "The End of SDS," 304–305.

[144] Tarrow, *Power in Movement*, 209.

[145] Ibid., 54, 127, 189–190.

[146] Miller, "The End of SDS," 306–307.

[147] Ibid., 307–308.

[148] Ibid.

[149] Jackson et al., "The Failure of an Incipient Social Movement," 40.

[150] Miller, "The End of SDS," 305.

[151] Christiansen, *Four Stages of Social Movements*, 4.

[152] Coy and Hedeen, "A Stage Model of Social Movement Co-Optation," 411, 413–426.

[153] John J. Macionis, *Sociology*, 9th ed. (Upper Saddle River, New Jersey: Prentice Hall, 2003), 619.

[154] Hopper, "The Revolutionary Process," 277.

[155] Edwards, *Natural History of Revolution*.

[156] Tarrow, *Power in Movement*, 206.

APPENDIX A. CODED ARIS CASE STUDIES

Key

- P = Preliminary state
- I = Incipient state
- C = Crisis state
- N = Institutional state
- R = Resolution state
- r = Radicalization
- i = Institutionalization
- p = Repression
- f = Facilitation
- s = Success
- l = Failure
- Six different manifestations of a failure state of resolution are possible and detailed in the text (factionalism, encapsulation, failure in preexisting network, failure to incorporate other leadership, lack of program for enthusiastic support, and highly publicized failures). Coding summaries should specify which form of failure applies to each relevant case.
- c = Co-optation
- m = Establishment with the mainstream
- e = Exhaustion
- > = Progression in phasing
- ^ = Reversal in phasing

Table A-1. Coded case studies with details.

Case Study	Analysis
Casebook on Insurgency and Revolutionary Warfare, volume I	
Revolution in Vietnam (1946–1954)	P > I > C > N > R(s)
	Preliminary state: 1885–1925. General unrest throughout early French occupation, characterized by repeatedly suppressed uprisings and protests, provided incubation for the emergent Vietnamese nationalist movement.
	Incipient state: 1925–1945. This stage was marked by formalization of the movement through creation of the Vietnamese Revolutionary Youth League in 1925 and others that followed. Discernible collective action and mobilization evolved through 1930 creation of the Vietnamese Communist Party and violent terrorist campaigns.
	Crisis state: August 19, 1945–September 23, 1945. The crisis state began with the outbreak of open revolution against French colonial rule after the end of Japanese occupation, ending with the French retaliation and ejection of the Democratic Republic of Vietnam.
	Institutional state: September 23, 1945–July 20, 1954. The Viet Minh survived French operations and waged guerrilla warfare immediately after reoccupation, later escalating into the prolonged First Indochina War.
	Resolution state: July 21, 1954. The Vietnamese Revolution resolved in a state of success with the International Geneva Conference, effectively ceding control of North Vietnam to the Viet Minh.
Indonesian Rebellion (1945–1949)	P > I > C ^ I > C > N > R(s)
	Preliminary state: 1908–1912. This state saw general discontent with Dutch occupation as socioreligious indigenous organizations critical of the Dutch formed.
	Incipient state: 1912–1927. The formation of political parties was later accompanied by peasant uprisings and demonstrations, providing the coalescence and formalization of ideas, institutions, and collective actions.
	Crisis state: 1927. A major Communist uprising against Dutch rule was staged and suppressed, defused by thousands of arrests and the exile of hundreds. Successful suppression of the uprising caused the nationalist movement to recede back into incipiency.

Case Study	Analysis
Indonesian Rebellion (1945–1949)	**Incipient state:** 1927–1945. The damaged nationalist movement later reasserted a formally organized presence in the 1935 formation of the Greater Indonesian Party, but it made no major moves for independence until Japanese occupation.
	Crisis state: August 1945–November 1945. Japanese defeat spurred popular mobilization, acquisition of arms, and rallies. This led to the declaration of independence on August 17, constituting a decisive confrontation with the legitimacy of post-World War II Dutch rule. The tactical British victory at the Battle of Surabaya marked the end of crisis, but the challenge to colonial authority by an emboldened resistance movement continued for years.
	Institutional state: November 1945–1949. Years of occupation and various military and diplomatic confrontations as the nationalist movement persisted.
	Resolution state: December 1949. The Netherlands succumbed to international pressure and transferred sovereignty to the Republic of the United States of Indonesia, constituting a resolution of the nationalist movement in success.
Revolution in Malaya (1948–1957)	$P > I > C \wedge I > C > R(e)$
	Preliminary state: 1930. Latent economic tensions under British rule were leveraged in the illegal foundation of the Malayan Communist Party (MCP). Ideas of Malayan independence milled via anti-imperialist Marxist ideological rationale.
	Incipient state: 1930–1941. The MCP provided formalized structure for organized collective action, generally resisting British economic influence, and after the 1937 Japanese invasion of China, Japanese imperialism with new Malayan Chinese recruits. Efforts included strikes and revolutionary committees, but MCP resistance was effectively suppressed by British operations until World War II.
	Crisis state: 1941–1945. The MCP and British High Command formed a wartime alliance as the Malayan People's Anti-Japanese Army to wage guerrilla war against occupying Japanese forces, using sabotage and ambushes in small units. The group survived Japanese reprisals to amass 4,500 soldiers by 1943 and received significant materiel support from allied forces in 1945. The crisis ended with the withdrawal of Japanese forces after World War II defeat.

Case Study	Analysis
Revolution in Malaya (1948–1957)	**Incipient state:** 1945–1948. Britain failed to demobilize MCP guerrilla units after the Japanese withdrawal, and unwillingness to allow Malayan independence resulted in the imposition of a federal system that discriminated against Chinese and Indian communities. These tensions, along with economic tensions, bolstered MCP mobilization against Britain for independence. Unrest amid the communist-led labor classes and hundreds of strikes gave way to terror and sabotage operations in 1948.
	Crisis state: 1948–1952. The MCP issued new orders for violent resistance to British rule, prompting violent riots. The British instituted "emergency" in efforts to suppress the MCP resistance, which used guerrilla tactics and sabotage. The crisis persisted through the 1951 peak of hostilities and assassination of Sir Henry Gurney.
	Resolution state: 1952–1960. Renewed British political, psychological, and military operations under the new high commissioner gradually succeeded in degrading MCP guerrilla forces and activities, exhausting the movement into obscurity by the 1957 allowance of Malayan independence within the Commonwealth. Britain finally ended the "emergency" in July 1960 despite limited guerrilla holdouts.
	P > I > C > N > R(s)
Guatemalan Revolution (1944)	**Preliminary state:** 1929–1941. The collapse of the Guatemalan economy in the Great Depression gave rise to high unemployment rates and general unrest among the working class, prompting elite support for the dictatorship of Jorge Ubico to prevent a potential uprising. Ubico's authoritarian measures and notably ruthless police force gradually compounded popular resentment of the regime. The United Fruit Company's accelerated land seizures and displacement of peasantry also contributed to the incubation of insecurity and unrest.
	Incipient state: 1941–June 1944. The outbreak of World War II caused increased economic turmoil and unrest in Guatemala, provoking still fiercer crackdowns by the Ubico regime. The failed May 1944 uprising in neighboring El Salvador resulted in a flow of exiled revolutionaries moving into Guatemala, coinciding with escalating university protests against Ubico in Guatemala City.

Case Study	Analysis
Guatemalan Revolution (1944)	**Crisis state:** June 1944–July 1944. Ubico suspended the constitution on June 22, 1944, sparking immediate backlash and calls for a general strike. Protesters gave Ubico an ultimatum, demanding reinstatement of the constitution. The Ubico regime responded with police firing into the protesters and a declaration of martial law, sparking a week of violence between the military and demonstrators as the revolt gained momentum. Ubico finally stepped down on July 1, 1944, marking the end of crisis-state confrontation.
	Institutional state: July 1944–November 1944. Federico Ponce Vaides, a member of the provisional junta established by the exiting Ubico, convinced the congress to appoint him interim president. Instead of implementing free elections, Vaides continued the suppression of dissent and intimidation of the indigenous population. Protesters and members of the military attempted violent insurrection or a coup but were brutally suppressed. Nevertheless, continued protests and demands eventually compelled the junta to declare elections.
	Resolution state: November 1944. Free elections resulted in the presidency of Juan José Arévalo, who garnered extensive support from elements of the revolutionary movement for a landslide victory.
Venezuelan Revolution (1945)	P > I > C > N > R(1—factionalism)
	Preliminary state: 1936–1941. After twenty-seven years of dictatorial rule under Juan Vicente Gómez, Eleazar López Contreras took control with a new constitution and numerous reforms but simultaneously became the subject of milling and growing opposition among the young left-wing intellectual class. Contreras chose Isaías Medina Angarita as his successor in 1940, and Angarita continued moderate rule and allowed for the establishment of opposition parties.
	Incipient state: 1941–1945. The Democratic Action (AD) party was founded in 1941 and became the largest opposition party, marking a coalescence of collective action and mobilization through strenuous campaigns against the new regime in the 1944 elections. After significant electoral losses that would effectively give Medina and his party the ability to choose the next presidential candidate without opposition, AD began to ally with young military officers who had their own separate grievances to plan a coup d'état.

Case Study	Analysis
Venezuelan Revolution (1945)	**Crisis state:** October 18–19, 1945. The Medina government was quickly overthrown by coordinated military rebellion among young army officers and the AD, who simultaneously organized popular dissent and propaganda efforts in support of the coup. The confrontation began on October 18, and although there was some fighting, the coup forces quickly gained control of the government buildings in Caracas and secured Medina's resignation on October 19, ending the crisis-level confrontation.
	Institutional state: October 1945–November 1948. Authority was transferred to a junta composed of coup officers and AD party members, who instituted new policy measures and organized new elections. The AD-led government ruled for only a short time and suppressed a series of rebellions by disaffected military members, but it soon fell to another coup in November 1948.
	Resolution state: November 24, 1948. The army's resurgence to political power marked the effective failure of the revolutionary movement, which was unable to depoliticize the military and effectively curtailed the social and economic policies of the revolutionary government. Although the AD party persisted in Venezuelan politics, the 1948 coup marks the resolution state of the widely recognized three-year period of El Trienio Adeco.
Argentine Revolution (1943)	P > I > C > N > R(s)
	Preliminary state: 1930–1942. The coup of 1930 began a period of right-wing regimes wrought with corruption scandals and internal power struggles. Despite some early moves toward fascism, opposition influence in the military and business class maintained a trend of moderate rule for ten years under Agustín Pedro Justo and Roberto Marcelino Ortiz. Opposition and nationalist discontent continued in circular interaction through this period but did not coalesce into an incipient state until the ascension of Ramón S. Castillo.

Case Study	Analysis
Argentine Revolution (1943)	**Incipient state:** 1942–1943. Vice President Castillo took office in 1942 because of President Ortiz's failing health, bringing with him fascist sympathies and isolationist policies. Around the same time, disaffected officers in the Argentine Army created the United Officers' Group (GOU) as a nationalist secret society against the Castillo regime, and they organized toward a coup d'état. When no opposition parties could offer a viable candidate to run against Castillo in the 1943 elections and failed to mobilize resistance, the GOU acted.
	Crisis state: June 3–4, 1943. After Catillo's early attempts to suppress the impending coup, the GOU and affiliated military officers marched on the capital and surrounded government buildings, establishing military rule in twenty-four hours with minimal resistance and bloodshed.
	Institutional state: June 5, 1943–1946. Despite internal power struggles, the military succeeded in concentrating political power in itself and trading power among a string of military dictators until the election of GOU member Colonel Juan Perón in 1946.
	Resolution state: June 4, 1946. Perón's election, despite facing some opposition from competing rivals within the junta, marked the success of the military regime to secure lasting political power.
Bolivian Revolution (1952)	$P > I > C \wedge I > C > N > R(e/s)$
	Preliminary state: 1935–1942. The end of the Chaco War between Bolivia and Paraguay left Bolivia in economic decline and uprooted much of the large indigenous Indian population through the draft, beginning the development of a more engaged social consciousness. Simultaneously, lower-middle-class workers grew in political power under and against regimes composed of a Spanish-speaking elite, oligarchic class.
	Incipient state: 1942–1949. In opposition to entrenched conservative power, the Movimiento Nacionalista Revolucionario (MNR) was founded in 1942, with the competing Bolivian Communist Party and others soon following. This constituted the formalized, collective mobilization of dissatisfied elements in Bolivia. Governing authorities sought to stem the growth of left-wing and labor movements during the sexenio (the six years preceding revolution) but failed in the face of economic decline and growing social unrest.

Case Study	Analysis
Bolivian Revolution (1952)	**Crisis state:** 1949. Unable to muster enough political support to take power legally, the MNR attempted a coup in 1949. Although it was unsuccessful, the escalated confrontation garnered increased support for the MNR among workers and some military officers. However, this did not constitute a transition to an institutional state, as the MNR continued to be politically marginalized and suppressed, despite its strengthened position.
	Incipient state: 1949–1952. The MNR emerged as the most popular political party in 1951 elections, but refusing to allow the MNR to take power, President Mamerto Urriolagoitía installed a military junta to rule with General Hugo Ballivián Rojas as president.
	Crisis state: April 1952. Led by the MNR, coordinated armed revolts erupted in all major Bolivian cities, clashing with progovernment troops. The army was eventually routed and took control of the government in a matter of days.
	Institutional state: 1952–1964. The MNR ruled for more than a decade and instituted many long-standing reforms that were never reversed.
	Resolution state: 1964. Factionalism in the MNR quickly decelerated the pace of reforms toward a state of exhaustion (despite long-standing success), and polarization of leadership saw the virtual destruction of the party in 1964.
Cuban Revolution (1953–1959)	$P > I > C \wedge I > C > N > R(s)$ **Preliminary state:** 1944–1952. The successors of dictator Fulgencio Batista saw a breakdown of police forces, courts, and public administration in Cuba. A general, unfocused sense of destabilization grew with increasing corruption and instances of violence between political factions.
	Incipient state: 1952–1953. Batista's violent return to power, combined with the suspension of the constitution and ruthless repression, prompted Fidel Castro to organize anti-Batista forces.
	Crisis state: July 26, 1953. Castro led his forces in a failed raid against the Moncada Army Barracks; he and the few survivors were arrested.

Case Study	Analysis
Cuban Revolution (1953–1959)	**Incipient:** 1953–1958. After his 1955 pardon, Castro and his contemporaries began preparing in Mexico for another attempt at revolution with training, money, and weapons. Castro's forces landed in December 1956, meeting an attack by Batista's forces. Despite heavy losses, Castro and his forces regained strength through recruits and continued to grow.
	Crisis state: May 1958. Batista ordered a major attack on the roughly three hundred revolutionaries in an attempt to decisively end the revolution, but the forces survived the confrontation and continued to build strength.
	Institutional state: July 1958–1959. Representing his July 26th Movement, Castro signed an agreement with other Cuban resistance organizations, strengthening the revolutionary forces to the point that they had an equal footing with government forces, making significant gains and capturing Santa Clara.
	Resolution state: January 1, 1959. Batista resigned and fled the country, leaving Castro and his revolutionary government successful in securing power.
Tunisian Revolution (1950–1954)	P > I > C > N > R(s)
	Preliminary state: 1890–1907. Tunisia was made a French protectorate in 1881, and radical publications critical of the status quo began to emerge by 1890, each of which was soon suppressed.
	Incipient state: 1907–1934. The Young Tunisians party was founded in 1907, constituting an organized manifestation of the movement. The Young Tunisians party actively conspired with the Ottoman Empire against French rule but was driven underground by 1912. The Destour party was founded in 1920 to argue for legal reforms (but not autonomy) and included some members of the Tunisian nationalist movement. Dissatisfaction with the goals of the party led to a rift and the foundation of a new nationalist-oriented party.
	Institutional state: 1934–1953. The Neo-Destour party was founded in 1934, constituting the formalized and legitimate platform for the Tunisian nationalist movement. Neo-Destour carried out a subversive campaign combining negotiations with the French, threats, and isolated campaigns of violence. The party grew its influence despite French efforts to subdue it.

89

Case Study	Analysis
Tunisian Revolution (1950–1954)	**Crisis state:** 1953–1956. Neo-Destour led an armed rebellion and campaign of violence, in coordination with sporadic negotiations with the French, until complete autonomy was secured.
	Resolution state: March 1956. The successful transfer of complete authority over Tunisia concluded in March 1956 and a new government was established under Neo-Destour.
Algerian Revolution (1954–1962)	$P > I > C \wedge I > C > N > R(s)$
	Preliminary state: 1881–1918. The initial period of French rule in Algeria, between 1831 and 1881, was marked by armed resistance. After 1881, the country was pacified, but considerable tension remained between the local Algerians and the French settlers, largely surrounding issues of land tenure. However, there was little violence or organized opposition.
	Incipient state: 1918–1945. At the end of World War I, a new episode of political ferment began, inspired by Algerian Muslim officers and soldiers who returned home after the war. The native Algerian elites demanded increased political rights, while the lower classes wanted improved economic opportunities. Three organizations were formed to oppose French rule. The Fédération des Élus Musulmans d'Algérie sought to improve conditions through the complete integration of Algeria into France and the extension of political equality to the Algerian Muslim population. In opposition to this organization were two Algerian movements seeking independence, the Étoile Norde Africaine, which focused on economic issues appealing to the lower classes, and the Association of Ulemas, which was a religious organization that stressed Islam and the Arabic language as unique factors that separated Algeria from France. After 1940, the Vichy regime increased the political restrictions in Algeria, banning all the nationalist movements. Calls by the free French for the Muslim community to support the Vichy regime prompted nationalists to demand improved rights in exchange for their support.

Case Study	Analysis
Algerian Revolution (1954–1962)	**Crisis state:** 1945–1946. The nationalist organizations had grown during World War II, and at the end of the war the French settlers cracked down on the Algerian nationalist movements in a reaction to what they thought was an attempted uprising after the instigations of some Muslim extremists. More than four thousand Muslim Algerians were killed, and the Parti du Peuple Algérien, which had evolved out of the Étoile Norde Africaine, was largely destroyed. The French settlers returned to their position of dominance over the Muslim population and resisted Paris's attempts to liberalize the colonial regime.
	Incipient state: 1946–1954. In 1946, the Parti du Peuple Algérien was reconstituted as a legal political party known as Mouvement pour le Triomphe des Libertés Démocratiques (MTLD), and the Fédération des Élus Musulmans d'Algérie was renamed the Union Démocratique du Manifeste Algérien, which continued to seek increased rights through integration with France. In 1948, the members of the MTLD formed a secret organization for the purpose of carrying out paramilitary actions against the government, leading to the creation of the Front de Libération Nationale (FLN), which began militant operation on October 31, 1954.
	Crisis state: 1954–1956. On the night of October 31, 1954, the FLN attacked French military posts throughout Algeria and, on Radio Cairo, announced the beginning of the Algerian War of Independence. The military was caught off guard, and the French military was forced to respond with heavy units meant to fight the Soviets because they did not have counterinsurgency forces in Europe or Algeria. Eventually forces from Indochina arrived in Algeria, and the military structure was adapted to fight the FLN guerrillas. The FLN started out with 2,000 to 3,000 fighters and was unknown to the Algerian people, but by the end of the war in 1962, it had 130,000. After the initial attacks on the military, the FLN retreated to rural areas to evade the French forces and build support. The group ambushed French patrols and attacked Algerian Muslims who were supporting the colonial government in an effort to reduce support for France. Terrorist attacks against French settlers required the French army to heavily guard the cities, preventing French forces from focusing on attacking the FLN forces in rural areas.

Case Study	Analysis
Algerian Revolution (1954–1962)	**Institutional state:** 1956–1962. In 1954, the FLN held the Soummam Valley Conference. The military structure was reformed, the rebel forces were named the Army of National Liberation, and a regular military command structure was established. In 1958, the FLN created the Provisional Government of the Algerian Republic, which included a premier, vice-premiers, and ministers for several key governmental departments. Provisional Government of the Algerian Republic decisions were binding on all members of the FLN. The Army of National Liberation continued fighting the French Army and, despite many military defeats, was able to increase its support among the people of Algeria as well as internationally. As the war dragged on, public opinion in France turned against the war. In 1958, the French Army in Algeria launched a virtual coup, bringing Charles de Gaulle back to power, in an attempt to prevent Paris from granting Algerian independence. Continued fighting in Algeria, however, forced de Gaulle to accept that self-determination for Algeria was a political necessity.
	Resolution state: 1962. On March 18, 1962, the French government and the FLN reached an agreement on a cease-fire and plans for an Algerian referendum on the future status of Algeria. The referendum was held on July 1 and returned an overwhelming majority for independence.
Revolution in French Cameroun (1956–1960)	P > I > C > N > R(p/f)
	Preliminary state: 1945–1951. Economic development during the colonial period created a middle class that was Western educated and largely detribalized. It consisted of traders, farmers, industrial workers, civil servants, students, and intellectuals. This group was antagonistic to both the colonial administration and the traditional African elites. Despite the middle class's growth and increasingly significant role in the administration of the colony, most of the economic benefits were going to Europeans and the indigenous traditional elites. After World War II, the new African middle class increasingly called for independence throughout France's African colonies. In 1947, the Rassemblement Democratique Africain (RDA), a leftist liberation organization spanning the French African territories, was created with the purpose of pushing for independence.

Case Study	Analysis
Revolution in French Cameroun (1956–1960)	**Incipient state:** 1951–1955. In 1951, the RDA ended its relationship with the French Communist Party, prompting the Camerounian branch of the RDA, the Union des Populations Camerounaises (UPC), to split from the organization and maintain its communist ties. Its demands of the French colonial authorities included unification of French Cameroun with the British Cameroons, establishment of an elected legislative assembly, establishment of a governing council with an African majority, and the fixing of a date for full independence. The group's activities during this period consisted of organizing demonstrations and distributing anti-French propaganda.
	Crisis state: 1955–1955. On April 22, 1955, a mass meeting of the UPC and communist-linked trade unions was followed by a series of violent anti-French demonstrations in Douala and Yaounde, during which some Europeans and many Africans lost their lives. In July, the French authorities outlawed the UPC.
	Institutional state: 1955–1960. The UPC continued operating as an illegal underground organization and carried out terrorist and paramilitary attacks on the French as well as on Camerounian officials working with the French. Many of the attacks consisted of sabotaging transportation and communications, burning plantations, and attacking villages and traditional leaders considered to be supporting the French. The UPC initially moved its headquarters to British Cameroons, then to Cairo, and finally to Conakry after Guinea became independent. The UPC refused to cooperate with elections organized by the French authorities in 1956 and called for a boycott. Thus, the UPC did not recognize the elected African leaders who were meant to lead Cameroun toward independence. The organization derived most of its support from the provinces of Sanaga-Maritime and Bamileke, as well as from urban residents from those two areas. Despite many of the fighters being members of the Bamileke tribe, the organization was not ethnically organized and included members of numerous ethnicities.

Case Study	Analysis
Revolution in French Cameroun (1956–1960)	In 1958, the elected government of Ahmadou Ahidjo offered amnesty to UPC members who returned to legal politics, and it called in military forces from other French colonies to suppress those who did not accept the amnesty offer. Many guerrillas laid down their arms immediately, and others continued to fight. Violence stopped in the cities and in Sanaga-Maritime Province but continued in Bamileke District. One of the UPC leaders who accepted Amnesty, Mayi Matip, created a legal party, the Force de Reconciliation, which represented the UPC demands through the existing political system.
	Resolution state: 1960. After Cameroun gained independence, much of the impetus for the UPC disappeared. The amnesty, which allowed the UPC's demands to be heard in the legal political process, also reduced the attractiveness of armed struggle. After 1960, some UPC fighters continued to operate, but their numbers were estimated at around two thousand. The attempted revolution was brought to an end through a combination of repression and facilitation, with many of the goals of the UPC met by the independence of Cameroun and the acceptance of former UPC members in the political process. Those elements that continued the militant struggle were repressed by the military.
	P > I > C > R(s)
Congolese Coup (1960)	**Preliminary state:** 1959–1960. On January 5, 1959, a riot broke out in Leopoldville (now known as Kinshasa), during which thirty thousand unemployed Congolese, led by members of the political elite, protested against economic conditions. The protests quickly took on political overtones, and the Congolese political parties demanded economic and social reforms and eventual independence from the Belgian colonial authorities. On January 13, the king of Belgium announced his intention to grant Congo independence, and in October the Belgian government announced plans for Congolese initial elections in December 1959 to select representatives to negotiate with the Belgian authorities, followed by independence in 1960. Another round of elections was carried out in early 1960 to create a government to take over from the colonial authorities, which resulted in a coalition government. The leaders of the two main parties, Patrice Lumumba and Joseph Kasavubu, became prime minster and president, respectively, despite a difference of opinion on whether Congo should be a centralized state or a looser federation. Independence was granted on June 30, 1960.

Case Study	Analysis
Congolese Coup (1960)	**Incipient state:** 1960. On July 8, 1960, elements of the Congolese National Army mutinied in Leopoldville against their white officers. The mutiny spread throughout the country and mob violence ensued. The riots resulted in a mass exodus of Belgian personnel from the Congo. Many of the Belgians were supposed to have stayed on as civil servants under the new government, and their absence limited the ability of the new government to continue functioning. The government dismissed white officers in the Congolese Army in an attempt to pacify the mutiny, and Colonel Joseph Mobutu (later known as Mobutu Sese Seko) was appointed chief of staff of the Congolese National Army. On July 11, Belgian paratroopers occupied the major cities, and the province of Katanga declared its independence. In August, Kasai Province also declared its independence from the central government. On July 12, Prime Minister Patrice Lumumba asked the United Nations (UN) for assistance in removing the Belgian forces and restoring order. The UN launched the first peacekeeping operation with significant military force, which first arrived on July 15. A UN resolution demanded the withdrawal of all Belgian forces. However, the UN forces refused to assist Lumumba in retaking Katanga and Kasai, and Lumumba's attempts to do so with only Congolese forces resulted in a power struggle. On September 5, 1960, Prime Minister Lumumba and President Kasavubu both ordered the dismissal of the other from office. The legislature refused to accept either dismissal. Without sufficient support from the UN to allow him to retake Katanga Province, Lumumba began accepting assistance from communist countries and openly considered asking the Soviet Union to intervene directly. The chaos in the government, the massacre of Baluba tribesmen in Kasai Province by government troops, and the increasing ties between Lumumba and the Communist bloc led Colonel Mobutu to begin planning a coup. He visited the American embassy frequently in the days leading up to the coup, although it is unclear what role the United States played in the planning or execution of the coup. It appears that the coup was planned very quickly, with Mobutu forming an agreement with the officers stationed in Leopoldville on September 14, the day before the coup.

Case Study	Analysis
Congolese Coup (1960)	**Crisis state:** 1960. On September 15, 1960, the chief of staff of the Congolese National Army, Colonel Joseph Mobutu (later known as Mobutu Sese Seko), launched a military coup. There was no resistance, and the coup was completed in twenty-four hours. He placed Premier Lumumba under house arrest, dismissed the cabinet, and created a council to serve as a caretaker government. President Joseph Kasa-vubu remained in office.
	Resolution state: 1960. Mobutu announced that the army would stay in power until January 1961. He expelled the representative of the Communist bloc out of Congo and threatened to demand the removal of UN troops if they did not cooperate with his government. Mobutu succeeded in imposing order in the provinces of Leopoldville and Equateur in the western portion of the Congo. The second in command of the Lumumba cabinet, Antoine Gizenga, created a rival government in the east of the country, which was recognized by the communist nations and several African states. Katanga was under the control of neither government and continued to assert its independence. On February 9, 1961, President Kasavubu and Mobutu declared a resumption of normal governance, and Mobutu surrendered his political powers.
	$P > I > C > R(s)$
Iraqi Coup (1936)	**Preliminary state:** 1932–1936. Between achieving full independence in 1932 to the coup in 1936, there were five prime ministers presiding over nine separate cabinets because politics was based on constantly shifting personal relationships. The political instability was driven primarily by personal interests in attaining power, rather than by policy disputes.
	Incipient state: 1936. In the summer of 1936, Hikmat Suleiman, a former cabinet member who was expelled because of his ties to the opposition, began conspiring with General Bakr Sidqi. It remained a personal conspiracy between these two individuals until October, when the chief of staff traveled to Turkey, leaving Bakr as acting chief of staff. At that time, other members of the military were informed of the plan, as was the political opposition, the Ahali group. Hikmat and Bakr were able to organize the coup in less than a week.

Case Study	Analysis
Iraqi Coup (1936)	**Crisis state:** 1936. On October 29, the army marched on Baghdad, and leaflets were dropped on the city calling on the king to dismiss the cabinet and replace it with one under the leadership of Hikmat Suleiman. The king was informed that the military intended to leave him in power as long as he replaced the cabinet. The only violence consisted of the assassination of the minister of defense and four bombs dropped on government buildings, injuring seven people, after an ultimatum to dismiss the government had passed.
	Resolution state: 1936. The king, after meeting with the cabinet and the British ambassador, refused to allow the prime minister to take action against the coup and instead accepted the coup's demands, allowing the formation of a government composed of the revolutionaries. The army entered the city at 5:00 p.m. on the day of the coup, and Hikmat and Bakr assembled a new cabinet the next morning.
Egyptian Coup (1952)	$P > I > C > R(s)$
	Preliminary state: 1922–1942. In 1922, Egypt was nominally granted independence, but British military forces and commercial interests remained in the country and exerted significant control over the government. In 1939, a secret society within the Egyptian military was created, dedicated to the removal of the British influence from Egypt. The young officers who created the society were frustrated by the government's failure to effectively oppose British imperialism and considered the monarchy to be ineffectual. They sought the creation of a republic free of the influence of the British and the monarchy.

Case Study	Analysis
Egyptian Coup (1952)	**Incipient state:** 1942–1951. In 1942, the secret society was reorganized, and a Central Committee was created to direct the clandestine cells. During World War II, the secret society made connections with the Muslim Brotherhood as well as the German Army. Increased British military presence and surveillance prevented any anti-British action from within the Egyptian Army during the war. In 1945, the secret society was reorganized again to create military and civilian branches. It began to issue circulars and engage in propaganda, while also preparing for potential violent actions. In 1947, British forces left most of Egypt, with the exception of the Canal Zone, but this did not satisfy the young officers' desire for change. The war in Palestine in 1948–1949 sent a shock through the Egyptian military, causing many officers to conclude that revolutionary change was necessary within Egypt to prevent such a defeat from recurring. In 1950, the secret society restructured itself again, this time with the specific purpose of intervening in the political process of the country, and the group officially became known as the Society of Free Officers. It created a Revolutionary Command Council and elected Gamal Abdul Nasser and Mohammed Naguib to leadership positions.
	Crisis state: 1951–1952. After the military defeat by Israel in 1949, the Wafd Party, which was in control of the government, responded to the people's increasingly anti-Western sentiment. In October 1951, the Wafd government abrogated the Anglo-Egyptian Treaty of 1936, which allowed the British to keep troops in the Canal Zone and Sudan. The British then refused to leave, leading to anti-Western demonstrations and riots in Cairo and Alexandria. In January 1952, during a major riot in Cairo, mobs attacked foreign, mostly British, targets. The Wafd government refused to use the police to prevent the riot, prompting the king to dismiss the government. Over the next six months, violence continued as a series of prime ministers appointed by the king were unable to impose order. The prestige of the king, which had been declining over time, dropped considerably because of his nondemocratic intervention in government. A coup was planned for March 1952 but was delayed after a key member backed out of the plan.

Case Study	Analysis
Egyptian Coup (1952)	Coup plans were reinstated after July 15, 1952, when King Farouk dissolved the executive committee of the Military Club, on which the Free Officers held a majority, after the election of General Naguib. Fearing that Naguib would be arrested, the Free Officers decided to launch the coup as soon as possible. They carried out the coup in only a few hours in the middle of the night on July 22–23, 1952. They arrested members of the High Command and occupied government and communications buildings. They stationed troops along the road to Suez to prevent a British intervention. Naquib was appointed the commander in chief, and Nasser, in control of the Revolutionary Command Council, assumed control of the government. A new politician was appointed on the 23rd to replace the former government.
	Resolution state: 1952. On July 26, King Farouk was expelled from the country, leaving the Revolutionary Command Council in complete control. In December, the 1923 constitution was abolished, and political parties were banned the next month. In 1956, after a power struggle within the Revolutionary Command Council, Nasser assumed complete control of the government, supplanting General Naguib. Later that year a new constitution was adopted.
	P > I > C > R(s)
Iranian Coup (1953)[a]	**Preliminary state:** 1941–1951. After British and Russian forces occupied Iran during World War II, Reza Shah was forced to abdicate, replaced by his twenty-one-year-old son, Mohammed Reza. The 1906 constitution, which Reza Shah had largely ignored, was implemented, shifting much power from the shah to the elected government. Throughout the 1940s, Mohammed Reza attempted to increase his powers and gained strong control over the military. Conflict between the shah and the elected government continued, especially after the British and Russian troops left Iran. The country was divided between supporters of the traditional elite, represented by religious leaders, regional landlords, and the shah, and supporters of reform, consisting of reformers seeking to reduce the powers of the shah, as well as communists, organized in the Tudeh Party, who sought more revolutionary change.

Case Study	Analysis
Iranian Coup (1953)[a]	**Incipient state:** 1951–1952. In May 1951, Mohammed Mosaddegh, a proponent of parliamentary government who supported reducing the powers of the shah, became prime minister. He immediately nationalized the Anglo-Iranian Oil Company and broke diplomatic relations with Britain. The British responded by blockading Iranian oil exports and using their influence in the world oil market to deter foreign oil companies from purchasing Iranian oil. The British appealed to the United States for assistance in pressuring Iran to accept an agreement short of full nationalization. The United Kingdom and the United States were concerned about guaranteeing access to Iranian oil supplies, as well as preventing increased communist influence in Iran. The British considered using military force to secure the oil facilities and began discussions with the United States about removing Mosaddegh from power through a coup. The United States dissuaded the British from taking action until negotiations had been exhausted.
	Crisis state: 1952–1953. In July 1952, the United States and the United Kingdom agreed that the only way to resolve the crisis in Iran was through a coup. This decision was in response to Mosaddegh's attempts to increase his power at the expense of the shah. After failing to pass a new election law that would have limited the influence of the rural conservative elites, Mosaddegh instead manipulated the next election, preventing voting in rural areas, which assured him a supportive parliament of urban representatives. Mosaddegh attempted to decrease royal control over the military, which the shah resisted. Mosaddegh countered by encouraging strikes and mass demonstrations in the street. After three days of violent protests, the shah backed down. When Mosaddegh's continued attempts to increase control over the military while weakening the shah's influence in government provoked opposition from parliament, he dissolved parliament and called for a referendum allowing him to act without parliamentary approval.

Case Study	Analysis
Iranian Coup (1953)[a]	In late 1952, the CIA and MI6 began to make plans for a military coup that would remove Mosaddegh from power and return the shah to prominence.[b] They made contact with military leaders, as well as religious leaders and bazaar merchants, and created a plan for mass demonstrations to coincide with the military seizing power. The shah agreed to cooperate with the plan after General Zahedi, who would lead the coup and appoint himself prime minister, presented the shah with a predated letter of resignation, accompanied by a CIA guarantee that the shah would be allowed to remain in power.[c]
	On August 13, 1953, the shah exercised his constitutional power to remove the prime minister from office and replaced him with General Zahedi. Mosaddegh refused to relinquish power and forced the shah to flee the country. On August 19, the army, in conjunction with crowds organized by religious leaders and bazaar merchants who were cooperating with the military, deposed Mosaddegh and replaced him with General Zahedi. After a tank battle between Zahedi's forces and a smaller force defending Mosaddegh, the military was able to occupy Tehran, bring the pro-Mosaddegh crowds under control, and restore order.
	Resolution state: August 19, 1953. General Zahedi returned from hiding and assumed the position of prime minister to which the shah had previously appointed him. He welcomed the shah back to the country and later resigned, in keeping with their earlier agreement. After the coup, the shah was able to increase his power over the government and continued to rule in an increasingly autocratic manner until the 1979 Islamic Revolution.
	P > I > C > R(s)
Iraqi Coup (1958)	**Preliminary state:** 1941–1954. In 1941, Nuri as-Said became prime minster and assembled a coalition of wealthy families and politicians, which was able to impose order on Iraq. This government was disliked for its authoritarianism and nepotism, as well as its pro-Western foreign policy. Public sentiment developed against the Western orientation of the government and against the monarchy. Pan-Arabism also played a role in stirring discontent with the government, and this sentiment increased greatly after the Egyptian revolution of 1952, which spurred antigovernment protests in Iraq.

Case Study	Analysis
Iraqi Coup (1958)	**Incipient state:** 1954–1958. In 1954, the prime minster, Nuri as-Said, banned all political parties, driving the opposition underground. Both ultranationalists and leftists from the banned parties began cooperating with the already-illegal Communist Party. In 1956, they were joined by the Baath party. These parties began coordinating to oppose the government, and planning specifically for the July coup began several months beforehand and involved members of the army working with the banned political parties. The Communist Party was informed of the coup briefly before it occurred.
	Crisis state: 1958. On July 14, 1958, after Nuri ordered two brigades to move to Jordan to join the twelve thousand Iraqi troops already there supporting the Jordanian monarchy, the troops instead occupied Baghdad, assassinated the king and royal family as well as Nuri as-Said, and announced that they were replacing the monarchy with a republican government.
	Resolution state: 1958. The revolutionary regime instituted a three-person Council of Sovereignty and a Council of Ministers. Iraq withdrew from the Arab Federation, which had linked Iraq with Jordan, and left the Baghdad Pact, which had linked Iraq with Britain militarily. The government established political relations with the communist states and moved away from a pro-Western orientation.
	P > I > R(s)
Sudan Coup (1958)	**Preliminary state:** 1956–1958. Sudan was granted independence from Britain in 1956 without any Sudanese liberation movement or nationalist organization developing. The two political parties that formed the new government at independence were organized along religious lines, representing two different Islamic sects, while the main opposition party was a secular urban party. A political divide developed between pro-Egyptian and pro-British elements within the Sudanese government, as Egypt, under Nasser, became the forefront of the anti-Western sentiment within the Middle East. This division became even more acute after the Iraqi revolution of 1958, which aligned Iraq with the Nasserist movement. The traditional elites in Sudan increasingly worried that the left-wing political opposition could launch a Nasserist revolution.

Case Study	Analysis
Sudan Coup (1958)	**Incipient state:** 1958. During the summer and fall of 1958, the Sudanese government became deadlocked, with the parliament split among three parties that refused to cooperate. By November, the parliamentary system had broken down and had become publicly discredited, and the traditional elites, who controlled the two parties that constituted the coalition government, feared a radical Nasserist revolution from the urban population. Talk of a socialist revolution spurred the traditional elites to begin planning a coup with military leaders in the latter half of 1958.
	Resolution state: 1958. On November 17, 1958, the military seized control of government buildings and communications centers in Khartoum and declared that it was suspending the constitution and replacing the parliamentary government with a military junta. It was a bloodless coup, and there was no opposition. The military carried out the coup with the knowledge and support of the leaders of the two main religious factions, which in turn controlled two of the three political parties. As such, it was an internal coup by the elite, who aimed to change the structure of government while still maintaining the traditional elites' control. To prevent a revolution, the new military government made some moves to appease the Nasserists, while still maintaining a pro-Western stance overall. Junior military officers unhappy with the lack of reform from the new government made several attempts at counter-coups in 1959, but all were suppressed by the military.
	P > I > C > R(s)
Korean Revolution (1960)	**Preliminary state:** 1952–1955. After the end of the Japanese occupation, a democratic government was created, but only one democratic election took place and that was in 1948. In 1952, President Syngman Rhee effectively destroyed all political opposition to his government and denied the Korean people their promised democratic rights, thereby increasing discontent.

Case Study	Analysis
Korean Revolution (1960)	**Incipient state:** 1955–1960. There was no clear organization behind these protests, and they appear to have been largely spontaneous. There was a growing organization of political opposition, however, in the form of the Democratic Party, which was created by members of the previously banned Democratic Nationalist Party. The Democratic Party was not able to accomplish much through the political process and did not organize the protests that led to the overthrow of the Rhee government, but it did present an organization that was capable of taking over power from Rhee's Liberal Party.[d]
	Crisis state: 1960. Protests began on February 8, 1960, when students protested against restrictions on opposition campaigning leading up to the March 15 election. Larger protests, with thousands of people, broke out on election day and continued into April in several provincial cities. Police brutality toward the protesters provoked larger protests, and on April 19, student-led protests began in Seoul. The crowds grew to one hundred thousand people, and protesters gathered in front of government buildings, with some attacking police stations. On April 25, professors led students to the National Assembly with a set of demands. On April 26, the US ambassador made a statement in support of the protesters.
	Resolution state: 1960. President Syngman Rhee resigned on April 27. An interim government was organized, and the constitution was amended, changing the government from a presidential system to a parliamentary one. New elections were held on July 29, with the former opposition, the Democratic Party, winning a majority. The new government began reforms demanded by the protesters, but a military coup replaced the new government in May 1961.[e]
Chinese Communist Revolution (1927–1949)	$P > I > C > N > R(s)$
	Preliminary state: 1902–1921. Dissatisfaction with the Chinese imperial government led to a revolutionary movement that succeeded in forcing the abdication of the emperor in 1911. A republican government was instituted, followed by anarchy as regional warlords and rebellions continually challenged the weak central government. Modernization and Western ideas encouraged changes in society. People were dissatisfied with the status quo, but communism had not yet risen to become an ideology for them to rally behind.

Case Study	Analysis
Chinese Communist Revolution (1927–1949)	**Incipient state:** 1921–1925. In 1921, the Chinese Communist Party formed, and in 1923 the revolutionary leader Sun Yat-sen accepted offers of assistance from the Soviet government, creating a national government, the Kuomintang, which was supported by the Communist Party. Chiang Kia-shek was appointed the head of the army.
	Crisis state: 1925–1928. After the death of Sun Yat-sen, a struggle for leadership erupted between Chiang Kia-shek and the Communist Party. In 1926, Chiang led a coup and took control of the government, and the Chinese Communist Party created a rival government.
	Institutional state: 1928–1949. Mao established the Chinese Soviet Republic in Juichin, while Chiang Kia-shek gained full control over the Kuomintang, in Nanking. The Communists and Nationalists both developed political parties and armies that essentially constituted two separate governments vying for control over China.
	Resolution state: 1949. The Communists pushed the Nationalists off the mainland, and the Kuomintang retreated to Taiwan. The Communist Party established the Chinese People's Republic. This can be considered a success but not a complete victory, as the Chinese Communist Party considered Taiwan an integral part of China. The Nationalist government still existed and was still recognized by many as the legitimate government of China. The Nationalist government in Taiwan continued to hold the UN Security Council seat for China.
	$P > I > C \wedge I > C > R(s)$
German Revolution (1933)	**Preliminary state:** 1918–1921. After the end of World War I and the signing of the Treaty of Versailles, there was considerable discontent in Germany. The public, and the army in particular, felt that they had been betrayed by the government in the peace negotiations, and Germans felt that the Western powers, and France in particular, was unduly punishing Germany. The Weimar Republic was plagued with instability, with parties on the left and the right challenging the political system itself, and two attempted revolutions threatened to overturn the government.

Case Study	Analysis
German Revolution (1933)	**Incipient state:** 1921–1923. Hitler became the leader of the National Socialist German Workers' Party, referred to as the Nazi Party, in 1921, and began mobilizing support to seize control of the Weimar Republic. He created a paramilitary organization, the Sturmabeitlung, known as the Brownshirts.
	Crisis state: 1923. In November 1923, the Nazis launched the Munich Putsch, an attempt to seize control of the Bavarian state government, with plans to then take Berlin. The army refused to cooperate and put down the revolt, arresting Hitler.
	Incipient state: 1923–1933. The Nazi Party responded to the failed Putsch by avoiding illegal methods of gaining power and focusing on gaining political power through the existing system. The party continued to maintain the Brownshirts and created a second paramilitary organization, the Schutzstaffel, or SS.
	Crisis state: 1933–1934. After becoming chancellor through legal means, Hitler increased his hold on power by denying other parties some electioneering rights and burning down the Reichstag. Using a combination of official state power and unofficial pressure, Hitler was able to pass the Enabling Act, which gave him extensive personal control over the government. Through the Nazi Party he dissolved other political parties, trade unions, and state governments.
	Resolution state: 1934. Hitler's rise to power was completed when he combined the positions of chancellor and president, making himself the head of state, head of government, and chief of the Army.
Spanish Revolution (1936)	$P > I > C > N > R(s)$
	Preliminary state: 1923–1936. The monarchy, military, and the Catholic Church had traditionally held power in Spain, but the growing middle class disrupted this order. In 1923, popular discontent forced the military to take power, with the knowledge of the king, to prevent the masses from overthrowing the government. By 1931, discontent had grown enough to force the king to resign. The new republic went through several changes of government, with power alternating between the left and right. Neither side was able to effectively govern, and politics became increasingly polarized. Dissatisfaction within the traditional elites, including the military, continued to grow during this period.

Case Study	Analysis
Spanish Revolution (1936)	**Incipient state:** 1934–1936. Two political parties, the Falange and the Carlists, began creating paramilitary bodies in 1934 in preparation for a potential military confrontation with the left-wing government. Within the military, preparations for a coup began in 1936. The victory of the left in the 1936 elections, followed by socialist agitation and strikes, convinced the parties on the right that a socialist takeover was imminent. The alliance of right-wing parties began discussing a coup with several generals in March 1936.
	Crisis state: 1936. The revolution began on July 17, with military units loyal to the Nationalist cause attempting to seize cities across Spain and in Spanish Morocco. Military units loyal to the Republican government, aided by armed trade union members, fought back.
	Institutional state: 1936–1939. By the end of July 1936, it was clear that the Nationalist forces would not be able to complete the revolution quickly, and the character of the conflict developed from a chaotic revolution to a civil war with clear lines of battle. General Franco took the Army of Africa from Morocco to southern Spain, where it became known as the Army of the South. General Mola commanded the Army of the North, in the northwest of the country. Portugal, Italy, and Germany supported the Nationalists, while international volunteers and the Soviet Union supported the Republican government. The Nationalists attempted to gain public support through a campaign of nationalism, stressing their support for the Catholic Church and the monarchy. They described the Republicans as "reds" and stressed the perils of communism.
	Resolution state: 1939. In March 1939, the Republican government disintegrated, and on March 28, 1939, the Nationalist Army entered Madrid and within days controlled all of Spain.
Hungarian Revolution (1956)	$P > I > C > R(p)$
	Preliminary state: 1944–1956. After World War II, the Hungarian people were subjected to forced collectivization and the institution of communist policies, which were unpopular with most Hungarians. Opposition political parties were persecuted and banned. Discontent grew among the population.

Case Study	Analysis
Hungarian Revolution (1956)	**Incipient state:** 1955–1956. The Hungarian revolution was highly spontaneous and had little organization before the revolution began. To the extent that there was an incipient state, it consisted of intellectuals inspired by the former Hungarian premier, Imre Nagy, who was expelled from the Communist Party in 1955. The Hungarian Writers Union and the Petofi Circle, two groups of intellectuals, began discussing opposition to the Hungarian government. When the protests began, they were initially started by university students.
	Crisis state: 1956. On October 23, students began protesting in public squares in Budapest, and within twenty-four hours, they were joined by factory workers and soldiers. Soon the intellectuals were calling for restraint, but the workers took the lead in the fighting. Hungarian police, aided by Soviet troops, fought with the protesters, but by October 30 the protesters had won and Imre Nagy had assumed the premiership and appointed a new cabinet. The new government declared the end of one-party rule and Hungarian neutrality but did not last long enough to implement any substantive changes.
	Resolution state: 1956. On November 4, Soviet troops returned and within a week had retaken control. A one-party communist regime replaced the revolutionary government, which had only just begun to function.
	P > I > C > R(s)
Czechoslovakian Coup (1948)	**Preliminary state:** 1943–1945. During World War II, the leaders of the Czech government in exile, disillusioned after the failure of France and Britain to support Czechoslovakia in the Munich Agreement, sought Russian support for a postwar Czech government. Many Czech exiles became supportive of communism during the war.
	Incipient state: 1945–1947. When Czechoslovakia was liberated from German occupation, the National Front was established by agreement of all of the anti-Nazi political parties. Anti-communist parties were banned from government, creating a left-wing government. The Communist Party gained key ministries in the 1946 election as part of a left-wing coalition government.

Case Study	Analysis
Czechoslovakian Coup (1948)	**Crisis state:** 1947–1948. The Soviet Union pressured the Czech government to reject aid from the Marshall Plan and coerced the government to drop a planned treaty with France. The minister of the interior, who was a communist, began replacing noncommunist police chiefs with communists. When ordered to stop and rescind the appointments, the minister of the interior refused, ignoring a legal order from the cabinet. When noncommunist parties, working with the president, used procedural rules to deadlock the government, the communist-controlled security forces entered the capital and demanded that the president replace the existing cabinet with one picked by the Communist Party. The president gave in to demands and accepted the new cabinet. The new cabinet assured that the 1948 elections listed only communist candidates.
	Resolution state: 1948. After the 1948 election, which gave the Communist Party complete control of the government, a Soviet-style constitution was adopted. Political opposition was neutralized and almost all businesses were nationalized. Czechoslovakia became a member of the Warsaw Pact and the Council of Mutual Economic Assistance, integrating it into the Soviet economy.

Casebook on Insurgency and Revolutionary Warfare, volume II

	$P > I > C > N \wedge I(a)$
NPA (1969–present)	**Preliminary state:** 1950–1969. The preliminary state saw widespread discontent with socioeconomic conditions and government corruption as well as lingering networks from the 1950s Hukbalahap (Huk) armed resistance. Mass student protests in the late 1960s led to the creation of the Communist Party of the Philippines in 1968. In 1965–1966, there were demonstrations in Manila against the presence of US military bases, which were viewed as symbols of repression.
	Incipient state: 1969–1971. The New People's Army (NPA) formed from remnants of Huk in 1969. Bernabe Buscayno and Jose Maria Sison emerged as leaders of the resistance, leading the NPA and the Communist Party of the Philippines, respectively. This period saw mobilization around the "Protracted People's War," development of a permanent physical base of operations, and growth of support base to include church and labor unions. Violent government crackdown on general opposition drew recruits to the resistance movement's cause and provided a common enemy in President Marcos.

Case Study	Analysis
NPA **(1969–present)**	**Crisis state:** 1971–September 22, 1972. NPA bombed a Liberal Party rally in 1971. The discovery of a shipment of arms destined for NPA from China led the government to declare martial law.
	Institutional state: September 22, 1972–1992. The NPA's support base swelled in response to the government's declaration of martial law. The government began military counterinsurgency tactics. NPA carried out offensive military campaigns from 1981 to 1985. During the mid-1980s, NPA controlled several governmental administrative agencies and operated in all seventy-three provinces. In 1986, NPA rejected the popular uprising that ousted Marco, and Aquino's election led to a decline in support at the height of NPA's organizational and military strength. Internal purges in 1988 led to a decline in NPA's public image and legitimacy. 1987–1990 included the highest numbers of NPA combatants, assassinations, and attacks, despite the group's continued decline in popular support and continued infighting. Professionalization of the Philippine army led to more sophisticated counterinsurgency tactics.
	Incipient state/abeyance: 1992–present. The government recognized the Communist Party of the Philippines as a legitimate political party. The NPA carried out the Second Great Rectification Movement from 1992 to 1998, eliminating those accused of being internal agents of the government. Growth and modernization of the economy, along with increased availability of democratic means of reform, continued to reduce support for NPA. Today, small NPA guerrilla units still operate, but there has been a major decline in encounters with government forces as well as in the group's recruitment and support.
FARC **(1966–present)**	P > I > C > N > R(i)
	Preliminary state: 1945–1966. The period of extreme political polarization known as The Violence legitimized political violence and led to development of several insurgent groups. The Revolutionary Armed Forces of Colombia (FARC) formed in 1964, and Ejército de Liberación Nacional (National Liberation Army, or ELN) formed in 1966. Polarization along Liberal/Conservative lines translated into a division between FARC(ELN) and the government during the 1960s, and competition for recruits and support continued among various insurgency organizations.

Case Study	Analysis
FARC (1966–present)	**Incipient state:** 1966–December 1990. FARC developed a support base among the rural working class, while the ELN built a base among university students and the church. A limited government response allowed continued political and military growth. By 1978, FARC had established a formal leadership structure. The group launched strategic plans such as the Strategic Plan for Taking Power (1982–1990) and the Bolivarian Campaign for the New Colombia (1990). By 1990, FARC was the most powerful insurgent force in the country.
	Crisis state: December 1990–February 1991. Working with ELN, FARC coordinated the largest insurgency operation in Colombian history, Operation Wasp. The government began to understand organizational and military capabilities of the insurgency but remained slow to respond.
	Institutional state: February 1991–2002. The FARC established formal training centers, a strategic leadership arm, and a military academy. In the early 1990s, the group began relationships with narcotraffickers, and these connections continue today. In 1998, FARC carried out attacks on the Colombian Army and antinarcotic forces followed by series of attacks against the government. The Pastrana administration (1998–2002) took a passive stance, allowing FARC to effectively control large areas in Southern Colombia, known as the Despeje.
	Resolution state (institutionalization): 2002. The election of President Uribe brought an increase in aggressive counterinsurgency, but FARC continued to operate. FARC served as a pseudo-government providing social, health, and educational services to civilians and collecting taxes from farmers and drug traffickers in controlled areas. Many recruits viewed joining FARC as a way to gain a steady job. As of 2010, FARC's annual revenue was approximately $900 million. As of this writing, FARC has become the Common Alternative revolutionary force political party, and Colombia struggles to implement the peace deal reached in November 2015 successfully..
Shining Path (1980–1992)	$P > I > C > N \wedge C > R(p)$ **Preliminary state:** 1962–1968. Abimael Guzman, a professor, organized an activist network that eventually developed into Shining Path. A history of racial tension, regional isolation, and economic marginalization in the Peruvian highlands produced an aggrieved population.

Case Study	Analysis
Shining Path (1980–1992)	**Incipient state:** 1968–1980. Shining Path was founded with Guzman as its leader. Economic depression began in 1975, hitting highland regions especially hard. Economic discontent mixed with racial tensions provided a breeding ground for Shining Path in the region. Shining Path leveraged university and community networks and began reaching out to the indigenous population to develop cultural connections and understanding. The group also developed a radical communist ideology, a membership indoctrination process, exclusive recruitment criteria, and a leadership structure.
	Crisis state: April/May 1980–1982. Guzman declared the start of armed struggle, and military attacks began. The government was slow to react, allowing the insurgency to operate unchallenged for two years. When the government did react, it was with excessive force, increasing the insurgents' legitimacy and popular support.
	Institutional state: 1982–1989. The success of military attacks by Shining Path and the harsh government response resulted in a rapid increase of support for the insurgency. The group continued violent attacks while effectively controlling the Ayacucho region. During this stage, Shining Path developed connections within the drug trafficking industry to increase its revenue. The government declared an emergency zone in the thirteen insurgency-controlled provinces in 1984, but this did not deter the insurgency's operations. The group strategically turned toward urban organization, expanding its support base, in the late 1980s.
	Crisis state: 1989–September 12, 1992. At the height of its military power, Shining Path began attacking urban centers. President Fujimori, elected in 1990, ramped up counter-insurgency efforts and economic aid to the highlands. By 1991, the population increasingly perceived Shining Path as a terrorist organization and a threat to the nation. Fujimori declared a state of emergency, suspended the constitution in April 1992, and declared a new military offensive in August.
	Resolution state (repression): September 12, 1992. Government intelligence forces captured more than a thousand high- and mid-level leaders, including Guzman. For the next year, the government confronted and repressed the remaining forces and successfully induced the captured Guzman to publicly denounce the violence and call for peace. By 1994, violence had been reduced to its lowest level since the insurgency began.

Case Study	Analysis
Iranian Revolution (1979)	P > I > C > R(s)
	Preliminary state: 1941–1977. Mohammed Reza Pahlavi was installed as shah by Iran's allies in 1941. A 1953 coup removed the democratically elected prime minister, replacing him with a shah-appointed prime minister. The 1960–1963 financial crises and electoral fraud produced large-scale riots and spurred opposition. In 1963, White Revolution reforms were enacted, causing unrest among the clergy and the merchant class, who disagreed with the shah's Western, secular reforms. Ayatollah Khomeini was arrested for speaking out against the shah and was exiled shortly thereafter. Opposition became increasingly vocal but remained unorganized and disconnected.
	Incipient state: 1977–December 1978. Khomeini successfully united the opposition into a cohesive mass. The National Front Party distributed open letters accusing the shah's regime of corruption and repression. A government-backed article denounced Khomeini, inciting protests that were brutally repressed by the government. Continued riots, strikes, and protests were met by security forces using harsh suppression tactics. In exile, Khomeini met with National Front officials, signaling unification of the opposition.
	Crisis state: December 1978–November 1979. In an attempt to quell protests, the shah appointed Shahpur Bakhtiar as prime minister, and Bakhtiar accepted on the condition that the shah leave the country. With the shah gone, Khomeini created the Council of the Islamic Revolution, returned to Iran as a hero, and appointed Mehdi Bazargan as prime minister.
	Resolution state (success): November 1979. The resistance declined as Khomeini solidified power over Iran, primarily through the Council of the Islamic Revolution and Hizbollah.
FMLN (1979–1992)	P > I > C > N > R(m)
	Preliminary state: 1931–1970. During this time, there were severe economic disparities between the landed elite and the peasant population. The Salvadoran Communist Party formed in 1930, followed by a military coup in 1931 and a ruthless military suppression of a peasant uprising, known as The Slaughter, in 1932. This period saw continued polarization, cultural breakdown, and dissolution of social ties after World War II.

Case Study	Analysis
FMLN **(1979–1992)**	**Incipient state:** 1970–March 1980. The Communist Party began negotiations with other communist insurgent groups in an effort to form a unified organization. Several left-wing insurgency groups formed throughout the 1970s. Negotiations in 1980 established the Farabundo Martí para la Liberación (FMLN) as the single revolutionary party.
	Crisis state: March 1980–1981. Archbishop Oscar Romero was murdered by a right-wing, presumably government-backed, death squad. The insurgency's ranks grew and there was increased demand for armed opposition. FMLN carried out its first major offensive, and the military responded by massacring a village. The violence escalated to civil war.
	Institutional state: 1981–January 16, 1992. FMLN created the Political-Diplomatic Commission to garner international support and recognition. The group functioned as an alternative government by providing assistance to peasant villages. In 1983–1984, FMLN shifted its strategy to more mobile warfare characterized by assassination, guerrilla tactics, and economic sabotage. Facing a stalemate and declining support both domestically and internationally, FMLN launched its second, and final, offensive in 1989. After this failed offensive, FMLN began to see military victory as unattainable, and peace negotiations began in 1991.
	Resolution state (establishment with mainstream): January 16, 1992. A peace accord was signed, with FMLN accepting the government's concessions and becoming a recognized political party. Under UN observation, FMLN and the Armed Forces of El Salvador demobilized.
KNLA **(1949–present)**	$P > I > C > N \wedge C > N$
	Preliminary state: 1853–1947. A long history of ethnic tensions weakened alliances and produced waves of conflict between the Burmans and the Karens, the two largest ethnic groups in Burma. The British colonizers exploited these tensions by favoring the Karens.
	Incipient state: 1947–January 1949. Various Karen organizations coalesced under the Karen National Union (KNU). The KNU issued a coherent demand for Karen autonomy, marked by peaceful demonstrations. The movement strengthened during the country's transition from colonial rule when the government's denial of an autonomous Karen state became clear. The insurgency armed itself to resist military incursions from the new Burmese government.

Case Study	Analysis
KNLA **(1949–present)**	**Crisis state:** January 1949–1962. The Burmese Army began military operations against the KNU, and a prolonged civil conflict ensued as various ethnic groups took up arms. A 1962 military coup led to the establishment of a new military government and increased repression of insurgency groups.
	Institutional state: 1962–March 1988. KNU persisted through the crisis state and, in 1975, the Karen National Liberation Army (KNLA) formed and became the armed branch of KNU. KNU and KNLA operated a quasi-government along the Thai–Burmese border and established a movement culture characterized by a flag, a coat of arms, national dress, a national anthem, and a history curriculum. During this period, KNLA saw its highest levels of troops as well as record profits from trade and taxation.
	Crisis state: March 1988–1989. KNLA launched a democratic uprising with other insurgent groups. Demonstrations continued throughout the spring and summer. The Burmese Army repressed the demonstrations, culminating in the killing of one thousand demonstrators on August 8, 1988. After a military coup in September 1988, the majority of ethnonationalist groups agreed to a crease-fire in 1989. KNLA did not accept the agreement and became the principal target of the Burmese Army.
	Institutional state: 1989–present. KNLA was strengthened by the crisis state and deepened its stance as an equal opposition player to the government. A temporary cease-fire in 2007 resulted in Burmese forces' withdrawal from designated border areas. Burmese forces launched a new offensive against KNLA in 2009. Today, KNU/KNLA receives external and domestic support, especially from the Karen population. As of this writing, Myanmar's government struggles to implement the National Cease-fire Agreement reached in October 2015, and the KNLA threatens nonparticipation in the Panglong Peace Conference. Skirmishes among ethnic groups and with the government continue..

Case Study	Analysis
LTTE (1976–2009)	P > I > C > N > R(p)
	Preliminary state: 1948–1972. Independence from British rule reintroduced a sense of distrust between the majority Sinhalese and minority Tamil populations. A brief period of cooperation ended after the 1956 election of the Sri Lanka Freedom Party running on a "Sinhalese-Only" platform. Both sides vocalized grievances, which were marked by Tamil antidiscrimination protests and anti-Tamil riots.
	Incipient state: 1972–July 1983. The Tamil New Tigers formed and in 1976 became the Liberation Tigers of Tamil Eelam (LTTE) under the leadership of Velupillai Prabhakaran. LTTE emerged as the foremost organization promoting Tamil self-determination, primarily through violent elimination of Tamil rivals. Anti-Tamil violence by Sinhalese civilians led to mass migration of Tamils to Tamil regions, spurring recruitment and support for the cause.
	Crisis state: July 1983–July 1987. Contention heightened after LTTE successfully ambushed a Sri Lankan army convoy. Large-scale anti-Tamil violence received minimal response from the government. LTTE established taxation, judicial institutions, nongovernmental organizations, and other social services in areas it controlled. LTTE conducted its first suicide bombing mission in 1987. India and the Sri Lankan government signed an accord, and India deployed military forces to Sri Lanka.
	Institutional state: July 1987–May 2009. In mutual opposition to Indian forces, LTTE and Sri Lanka cooperated to force the Indian military to withdraw in 1990. Cooperation quickly faded, and LTTE launched a string of political assassinations and suicide bombings. In 1997, the United States placed LTTE on its list of terrorist organizations, decreasing the group's flow of and access to external funds. Despite a 2002 cease-fire and a growing Sri Lankan army, violent confrontations continued. In 2006, Sri Lanka, with the support of rival Tamil groups, began a military campaign against LTTE, marked by human rights violations and large numbers of displaced citizens.
	Resolution state (repression): May 2009. The Sri Lankan army conducted an offensive, after which the Sri Lankan government declared final victory and the LTTE admitted defeat, resulting in resolution of the resistance through repression.

Case Study	Analysis
PLO **(1964–present)**	P > I > C ^ I > C > N > R(i) **Preliminary state:** 1917–1964. This period saw the reemergence of ethnic and religious tensions and insecurity when the British took control of Palestine and the 1917 Belfour Declaration encouraged mass migration of Jews to the area. General unrest among Palestinians in reaction to Jewish settlers culminated in the Arab Revolt from 1936 to 1939, which was suppressed by British forces. After World War II, the UN reorganized the territory into Arab and Israeli. Israel declared a Jewish state on May 15, 1948. The Palestinian population dispersed to Gaza and the West Bank. **Incipient state:** 1964–January 1965. The resistance coalesced under the Palestine Liberation Organization (PLO), set up as an umbrella organization for various pro-Palestinian groups. PLO established a single Palestinian army with the unified goal of fighting Israel, as well as a Palestinian parliament, national council, and treasury. **Crisis state:** January 1965–June 1967. PLO launched its first military operations, but the majority were unsuccessful. PLO developed training bases in Jordan and Syria. In June 1967, Israel captured several disputed territories and gained effective control of Palestinian land after its victory in the Six-Day War. **Incipient state:** June 1967–February 1968. PLO slipped back into an incipient state after facing humiliating defeat, loss of territory, and the public's frustration with the movement. Yasser Arafat, leader of Fatah, took over as leader of PLO. Despite organizational shifts under Arafat, the PLO continued to struggle with internal factionalism because of its lack of central authority, cohesive identity, and clear strategy. **Crisis state:** February 1968–1987. PLO success in the Battle of Karameh brought increased confrontation. PLO shifted its focus to armed resistance and terrorism, rather than diplomacy, as its dominant strategy. PLO carried out attacks against Jews outside of Israel, including hijacking several passenger planes in 1970. In 1974, PLO began attacks on Israel from Lebanon. When the First Intifada erupted in 1987, PLO shifted its focus to demonstrating its ability to rule, and the group withdrew from fighting. Hamas then stepped in as the key Palestinian fighting force.

Case Study	Analysis
PLO **(1964–present)**	**Institutional state:** 1987–January 2006. PLO persisted through the crisis state but then shifted its strategy. Bureaucratization of the PLO after the Palestine National Council, led by Arafat, recognized the 1947 UN resolution establishing two states. PLO/Fatah acted as a makeshift government. In the 1993 Oslo Accords, PLO and Israel recognized one another and outlined steps toward Palestinian self-rule. Israel gradually transferred some control to the Palestinian Authority. Yet, demands for a more permanent solution led to the violence of the Second Intifada from 2000 to 2003. A cease-fire in 2005 led to the withdrawal of Israeli troops from the Gaza Strip and some West Bank settlements.
	Resolution state (institutionalization): January 2006. PLO moderated its tactics, leading to its decline as the more radical Hamas took control of the movement, marked by its takeover of the Palestinian Authority in 2006 elections. Numerous PLO supporters switched allegiance to Hamas. In 2010, PLO agreed to US-mediated talks with Israel. As of this writing, PLO remains in operation but largely through institutional channels rather than armed resistance. The recent removal of the PLO secretary-general sparked rumors of internal factionalism.
Hutu–Tutsi Genocides (1994)	$P > I > C \wedge I > C > R(f)$
	Preliminary state: 1885–1957. This period saw ethnic tensions between the Hutu and the Tutsi, which were the result of colonial exploitation and institutionalization of the Tutsi hierarchy in Rwanda and Burundi.
	Incipient state: 1957–May 1965. Division between Hutus and Tutsis expanded and both sides, separately, developed a clear sense of in-group shared purpose. Western favoritism of Tutsis declined, and Belgian authorities granted Hutus a degree of political and social power. Hutu intellectuals wrote the Hutu Manifesto, advancing group consciousness and the social myth of Tutsi invaders. The Hutu Revolution in 1960 put Hutus in control of the new independent government, increasing the sense of insecurity and urgency among the Tutsi.

Case Study	Analysis
	Crisis State: May 1965–August 1972. Action breaks out in Burundi after the Tutsi government brutally repressed large-scale Hutu revolts. Years of political coups and assassinations from both sides culminated in the Tutsi government's slaughter of Hutus in Burundi from May to July 1972. In Rwanda, refugee Tutsis, known as *inyenzi*, began attacks on the Hutu nationalist government.
	Incipient state: August 1972–October 1990. Despite relative calm and perceived peace, a growing sense of victimization emerged on both sides, along with calls to action to combat the threat posed by the other side. The group Akazu, or "Hutu Power," started popular radio and news campaigns to perpetuate the image of the evil, threatening, and alien Tutsi. Meanwhile, Tutsi refugees in Uganda formed the Rwandan Patriotic Front (RPF) with the shared purpose of forcing Tutsis' return to Rwanda.
Hutu–Tutsi Genocides (1994)	**Crisis state:** October 1990–July 1994. Civil war broke out after the RPF launched its first attacks. After limited success, the RPF captured territory in 1992. After being made more vulnerable, the Hutu president agreed to peace negotiations. In response to peace accords, Akazu formed armed paramilitary organizations and carried out attacks from 1992–1993. On April 6, 1994, the Hutu president was assassinated and days later Akazu seized power and carried out the genocide of Rwandan Tutsis. The RPF ended the genocide by taking Kigali in July 1994.
	Resolution state (facilitation): July 1994. The genocidal violence was resolved when RPF declared victory and took control of Rwanda. Vested interests brought about the decline of conflict through repression and concessions. This state saw a degree of institutionalization and power sharing between Hutu and Tutsi groups, but conflict continued in Burundi and the Democratic Republic of the Congo.

Case Study	Analysis
KLA (1996–1999)	P > I > C > R(s)
	Preliminary state: 1878–1993. This period saw the emergence of Albanian nationalist aspirations to gain autonomy. Control of Kosovo changed hands several times, but the region was often effectually ruled by Serbia. Political and economic repression by the Serbian state led Albanians to establish a shadow state in Kosovo. From 1968 to 1993, Albanian nationalism and demands for Kosovar self-determination surged, characterized by riots and student protests met by repressive Serbian responses.
	Incipient state: 1993–March 1998. During this time, there was discernible collective action, a shared sense of a problem, and a common attribution of blame for that problem. Various military and political organizations formed, including the Kosovo Liberation Army (KLA), to represent Albanian Kosovars. Movement leaders, known as the "group of four," emerged, networks of secret cells developed, and KLA's goal of armed resistance began to overshadow visions of passive resistance and diplomacy. Milosevic became president of Serbia in 1989 on a nationalist platform, and he abolished Kosovo's autonomous status. The movement coalesced by the 1990s, marked by independent Kosovar elections and the establishment of a parallel system of government.
	Crisis state: March 1998–June 1999. Resistance action escalated and confrontations heightened in response to the murder of the Kosovar liberation hero Adem Jashari at the hands of Serbian paramilitary. The resistance formalized during this time, characterized by KLA uniforms, its establishment of political and military leaders, and controlled media campaigns. A NATO air campaign commenced in March 1999.
	Resolution state (success): June 1999. Serbian President Milosevic agreed to a peace accord, and a UN resolution established a UN interim administration and called for deployment of peacekeeping forces. The KLA demobilized and re-formed as a political party, with KLA leaders becoming political leaders.

Case Study	Analysis
PIRA (1969–2001)	P > I > C > N > R(f)
	Preliminary state: 1912–1969. Sectarian tensions arose between Catholics and Protestants in Northern Ireland because of disparate economic, social, and political opportunities. Catholics vocalized their grievances in Irish Republican Army (IRA) campaigns throughout 1910–1920s and later civil rights protests in 1968–1969.
	Incipient state: 1969–January 1972. Collective action mobilized around Catholic grievances, marked by organized marches in Belfast and Derry. Violence erupted between 'protesters and police, and ultimately British troops, with minimal IRA involvement. The Provisional Irish Republican Army (PIRA) split from IRA with goal of returning the movement to armed resistance, and the group immediately began recruiting, strategic planning, mobilizing, and safeguarding Catholic neighborhoods.
	Crisis state: January–July 1972. Action broke out and contention heightened after British troops killed civilians in what is known as Bloody Sunday. PIRA began a bombing campaign in response. Unsuccessful secret talks took place between PIRA and the British secretary of state. PIRA bombs exploded across Belfast in what is known as Bloody Friday, resulting in nine civilian deaths.
	Institutional state: July 1972–April 1998. Despite fallout after Bloody Friday, the resistance persisted through the crisis state. PIRA shifted its focus to targets outside of Northern Ireland. PIRA continued operations characterized by waves of cease-fires, assassinations, bombings, hunger strikes, and protests.
	Resolution state (facilitation): April 10, 1998. The resistance declined after the Good Friday Agreement satisfied some of its demands. The agreement called for PIRA to disarm, enacted policing reforms, and set up power-sharing institutions. Sinn Féin, PIRA's political arm, became one of largest parties in Northern Ireland. Despite the movement's achievements, its primary goal of independence and unified Ireland was not met. In 2005, international observers announced PIRA's complete demobilization.

Case Study	Analysis
	P > I > C > N > R(s)
	Preliminary state: 1973–April 1978. Political and social grievances emerged after a coup established a Soviet-influenced government. Opposition groups emerged and vocalized preferences for political Islam and the removal of the foreign, imposed regime.
	Incipient state: April 1978–December 1979. The Saur Revolution brought a communist government to power. Disparate factions in society began to coalesce in response to the new government's reforms, which were seen as direct challenges to Afghan and Islamic customs, and the severe repression of opposition. The Soviet Union invaded Kabul and replaced the prime minister. The Soviet invasion produced a clear sense of a common enemy. Reactionary revolts led by Mujahidin forces spread across the nation.
Afghan Mujahidin (1979–1989)	**Crisis state:** December 1979–1984. The Mujahidin began guerrilla attacks and soon controlled most roadways, weakening the government. Soviet attacks in response increased in intensity.
	Institutional state: 1984–1990. Mujahidin forces persisted through the crisis state and gained strength via substantial increases in foreign-supplied aid, weaponry, and training. Despite internal factionalism, considerable external funding allowed the Mujahidin to put Soviets on the defensive by 1987. By 1988, the Mujahidin controlled Eastern Afghanistan, and the Soviets began to withdraw troops.
	Resolution state (success): 1990. The Mujahidin gained effective rule and instituted an Islamic government. The transfer of power fueled factional conflict, and by 1996 the Taliban controlled the government.

Case Study	Analysis
Viet Cong (1954–1976)	P > I > C > N ^ C ^ I > R(s)
	Preliminary state: 1954–1957. North and South Vietnam divided after the Geneva Accords ended the Indochina War. The communist Viet Minh controlled North Vietnam, and a US-backed capitalist regime controlled South Vietnam. Numerous Viet Minh units stayed behind in South Vietnam, laying the foundation for the future Viet Cong insurgency.
	Incipient state: 1957–1959. The movement's ideas and actions coalesced during this stage, marked by the start of the Viet Cong's assassination campaign, strategic recruitment in villages, and its propaganda in South Vietnam. High levels of government repression and corruption produced anger and distrust among South Vietnam civilians, providing the Viet Cong with a captive, willing audience.
	Crisis state: 1959–1964. The formalization of the resistance was signaled by North Vietnamese leader Ho Chi Minh's announcement of armed revolution against South Vietnam, North Vietnam's establishment of a Central Office of South Vietnam to oversee Viet Cong operations, and the creation of a political arm for the Viet Cong, the National Liberation Front (NLF). A coup in December 1963 signaled the vulnerability of the South Vietnam regime.
	Institutional state: 1964–January 1968. The Viet Cong was strengthened by its first battle victory against US forces and the political vacuum the coup created in South Vietnam during the crisis state. The movement solidified its power through continued recruitment, military operations, resource mobilization, and propaganda campaigns in South Vietnam.
	Crisis State: January 1968–March 1968. The movement returned to the crisis state after heightened confrontation during the Viet Cong-led Tet Offensive.
	Incipient state: March 1968–July 1976. The Viet Cong receded into incipiency after suffering heavy casualties during the Tet Offensive. The People's Army of Vietnam forces replaced those the Viet Cong lost, estimated at half the Viet Cong's fighting force. The Viet Cong ceased to be a military organization for the remainder of the war.
	Resolution state (success): July 1976. North and South Vietnam reunified under a communist regime, signaling the movement's resolution in success.

Case Study	Analysis
Chechen Revolution (1991–2002)	P > I > C > N > R(r)
	Preliminary state: 1919–1991. The historical experience with a unified nineteenth-century Islamic imamate left the group with latent collective consciousness. An unorganized restlessness emerged as the Red Army entered Chechnya as an occupier, and this restlessness was marked by uprisings throughout 1920–1930s. An aggrieved population and a common enemy existed based on discrimination, deportation, and exploitation under the Soviet regime.
	Incipient state: 1991–December 1994. Marked by the declaration of an independent Republic of Chechnya, the resistance coalesced in response to the fall of the Soviet Union. Chechnya was a de facto state without interference from a preoccupied Russia.
	Crisis State: December 1994–1996. Action broke out and confrontation heightened when Russian troops entered the Chechen capital, Grozny, resulting in the First Chechen War. The war ended in a cease-fire in August 1996.
	Institutional state: 1996–2002. Resistance persisted through the crisis state, and the group began renewed attacks shortly after the cease-fire, marked by kidnappings, terrorist attacks, hostage situations, and general lawlessness. The resistance established sharia law, Islamic courts, and training camps in Chechnya. Under the new leadership of Putin, Russian forces reentered Chechnya in 1999. Grozny was captured, and Putin declared direct rule from Moscow.
	Resolution state (radicalization): 2002. The resistance declined as its radical Islamist shift distanced the group from its initial nationalist-separatist position and demands. Increased Russian state opposition caused regionalization, dispersion of resources, and exacerbation of internal cleavages. The group's decline was also due to Putin's hard-line rhetoric describing members of the resistance as terrorists, highlighting the gap between the resistance and the Chechen people. As of this writing, there are signs of improved relations, but skeptical international observers maintain that relations are defined by repression and that the insurgency is still active.

Case Study	Analysis
Hizbollah (1982–2009)	P > I > C > N
	Preliminary state: 1943–1982. Lebanon declared independence in 1943 and instituted political power-sharing along religious lines. During this time, high levels of domestic sectarian violence were marked by the onset of civil war in 1975. The Shia "awakening" in the 1970s was characterized by the development of Shia militias and militant ideology in opposition to perceived religious, territorial, and political occupation and repression. Unrest and insecurity among the Lebanese population increased after the Israeli invasions in 1978.
	Incipient state: 1982–1992. Discernible collective action against Israeli targets, characterized by hit-and-run attacks and suicide bombings, intensified after the Israeli invasion in 1982. Hizbollah coalesced into a unified organization and issued its founding manifesto in 1985, articulating coherent goals and strategies, and the group established a political and a military arm.
	Crisis state: 1992–July 1993. Formalization of the resistance was signaled by Hizbollah's participation in the 1992 Lebanese elections. The resistance movement's actions escalated, leading Israel to launch the Seven-Day War (Operation Accountability) in retaliation. The public increasingly perceived Hizbollah as a provisional authority and a legitimate representative of the Shia population.
	Institutional state: July 1993–present. Hizbollah persisted through the crisis state and transformed into an equal opposition player with broadened appeal. The group's gains in domestic support were signaled by electoral victories in 2009. Hizbollah controls several media outlets, including a satellite channel and several radio stations and newspapers. Continued confrontation with Israel is characterized by cyberattacks, rocket launches, terrorist activity, and an all-out war from 2006 to 2008. The group's more recent actions include involvement in the Syrian civil war, fighting with Assad against Sunni Syrian rebels, as well as in domestic political conflicts in 2011, 2013, and 2014.

Case Study	Analysis
Hizbul Mujahideen (1989–present)	P > I > C > N > R(r)
	Preliminary state: 1947–1987. A sense of Kashmir self-determination emerged after India and Pakistan gained independence from British rule. The Kashmir population's grievances were related to perceived religious oppression, dismal economic prospects, and government corruption. This sense of insecurity was exacerbated by ongoing Kashmir territorial disputes between India and Pakistan.
	Incipient state: 1987–September 1989. This phase saw discernible collective action and mobilization around the ethnoterritorial demand for Kashmir's independence from India. Fraudulent elections in 1987 contributed to the group's mobilization and its view of violent insurgency as its only option. Hizbul Mujahideen emerged in 1989, heavily influenced by the success of the Afghan Mujahidin and backed by Pakistan's intelligence services.
	Crisis state: September 1989–1991. Action escalated as strikes, mass demonstrations, bombings, arsons, and political assassinations dismantled Kashmir's civil administration. Confrontation heightened as the Indian government ramped up counterinsurgency efforts with little regard for the civilian population. Low-intensity conflict transformed into a full-blown revolt.
	Institutional state: 1991–2000. Hizbul Mujahideen strengthened during the crisis state and established itself as the predominant insurgency movement on the ground in Kashmir. The group established several "liberated zones" throughout Kashmir. In 1995, radical foreign jihadists and Wahhabi organizations joined the insurgency and reframed the conflict as a religious war, leading to a gradual decline in domestic support for Hizbul Mujahideen.
	Resolution state (radicalization): 2000. The resistance declined when radicalized demands of jihad resulted in decreasing popular support for the group and increasing internal fragmentation. The organization remains active, marked by a public call for renewed jihad against India in 2008, but it is no longer the predominant actor.
EIJ (1928-2001)	P > I > C > N > R(l-encapsulation)
	Preliminary state: 1928–1975. Restless Islamic elites grew dissatisfied with an increasingly secular Egyptian regime. The spread of the Qutbist doctrine in the 1950s–1960s and Islamic student movements in the 1970s characterized the growth of a shared Islamic identity and purpose.

Case Study	Analysis
EIJ (1928-2001)	**Incipient state:** 1975–October 1981. An intellectual cadre developed and numerous underground groups coalesced around their shared goal of Islamic revolution and their common enemy in the Egyptian government. Several small jihadi cells coalesced into the Egyptian Islamic Jihad (EIJ) under the leadership of Muhammad Abdel Salam Farraj. Anger spread after the mass imprisonment of Islamic leaders in 1981.
	Crisis state: October 1981. Violent action broke out when President Sadat was assassinated by an army lieutenant with connections to the EIJ. Several days of violence were followed by the election of Hosni Mubarak and the swift government crackdown on Islamic groups.
	Institutional state: October 1981–2001. The insurgency persisted through the crisis stage only slightly weakened, but it regrouped by the mid-1980s. Through the 1980s–1990s, EIJ carried out terrorist attacks in Egypt and later against US targets and participated in regional jihad, particularly in Afghanistan. In 1995, EIJ announced that it had run out of funds, leading to reliance on funding from Al Qaeda.
	Resolution state (failure – encapsulation): 2001. The resistance organization declined because of dwindling membership and popular support in Egypt. Decline was also due to the group's limited organizational capacity in the face of increasingly effective counterinsurgency. The decline in mobilization and capabilities led EIJ to merge with Al Qaeda in June 2001, signaling EIJ's failure through encapsulation.
	$P > I > C > N \wedge C > N$
Taliban (1994–2009)	**Preliminary state:** 1990–1994. High levels of insecurity, violence, and fractionalization emerged after the Mujahidin took control of Afghanistan. Decentralized authority brought about disparate regional power holders. The political vacuum and dependence on poppy cultivation produced dismal economic conditions and high levels of criminal activity. An aggrieved Pashtun population emerged as a result of the Tajik-led Mujahidin.

Case Study	Analysis
Taliban (1994–2009)	**Incipient state:** 1994–1996. The resistance coalesced to form the Taliban, an offshoot of the fundamentalist Mujahidin faction. Discernible collective action was demonstrated as the Taliban moved through southern Afghanistan, swiftly taking control of territory. Desire for order and security after years of conflict allowed the Taliban to gain control with popular support.
	Crisis state: December 1995–September 1996. Action broke out as the Taliban entered northern provinces and began attacks on the government in Kabul. The Taliban seized Kabul from the Mujahidin and publicly killed a prominent Mujahidin leader.
	Institutional state: 1996–November 2001. The movement bureaucratized with the establishment of the Taliban government. Confrontations and territorial disputes with Mujahidin forces and regional militias continued, but the Taliban effectively ruled most of Afghanistan.
	Crisis state: November 2001–December 2001. US-led coalition forces invaded Afghanistan, and the Taliban government fell.
	Institutional state: December 2001–present. The Taliban persisted through the crisis state, fled to Pakistan, reorganized under the leadership of Mullah Omar, and launched renewed military campaigns in 2003. Throughout the 2000s, the Taliban engaged in confrontations with Afghan and US military forces. With the withdrawal of US troops in 2013–2014, the Taliban regained control of some territories. In July 2015, peace talks between the Taliban and the Afghan government were postponed when Omar was announced dead, possibly since 2013, and a new leader was chosen.
Al Qaeda (1988–2001)	$P > I > C > N \wedge C \wedge I$
	Preliminary state: 1978–1985. The Soviet invasion of Afghanistan bred high levels of discontent among the Islamic population. Political repression pushed social organization into religious institutions, encouraging a religious mobilization of the opposition. Growing Islamic activism in the region, US foreign policy, and the continuing Israeli–Palestinian conflict provided a consistent message of insecurity and oppression among Arab communities as well as a common enemy in the United States and US-backed regimes.

128

Case Study	Analysis
Al Qaeda (1988–2001)	**Incipient state:** 1985–1992. Discernible collective action and mobilization was marked by the formation of the Afghan Services Bureau (Maktab al-Khidamat) to recruit and train Muslims for anti-Soviet resistance in Afghanistan. A resistance leader and intellectual cadre emerged under Osama bin Laden. A social myth and religious narrative that resonated with the civilian population developed.
	Crisis State: 1992–1993. Action was signaled by the first Al Qaeda-attributed attack. With the World Trade Center bombing, Al Qaeda attacked outside of the Middle East.
	Institutional state: 1993–September 2001. The resistance was strengthened by the crisis state. Bin Laden returned to Afghanistan and developed close ties to Taliban and EIJ leaders. Organizational and strategic power was illustrated by the group's declaration of war against American (Western) occupation in 1996. Al Qaeda operated more as a decentralized network than as an organization, with presence in an estimated sixty countries by 2001.
	Crisis state: September 2001–October 2001. Action and confrontation escalated after Al Qaeda attacked the United States in September 2001. After the attack, the United States declared the War on Terrorism, with Al Qaeda as its primary target.
	Incipient state: October 2001–present. US and coalition forces launched military operations after the 9/11 attacks, driving Al Qaeda out of Afghanistan and into isolated regions in Pakistan. These events signaled the group's return to the incipient state. Al Qaeda became less of an organization and more of an operational philosophy providing strategy and resources for active affiliated groups around the world. US forces killed bin Laden in Pakistan in May 2011.
MEND (2005–2010)	$P > I > C > N > R(c)$
	Preliminary state: 1960–1966. The oil industry's rapid expansion after Nigeria gained independence in 1960 exaggerated preexisting socioeconomic, ethnic, and political disparities. Unorganized demands for control over oil reserves emerged, especially among the minority Ijaw population.

Case Study	Analysis
MEND (2005–2010)	**Incipient state:** 1966–December 2005. An armed insurrection against the government signaled the first discernible collective resistance. Several armed militia groups formed to fight the corrupt government and exploitative oil industry. Democratization in 1999 opened channels for government opposition to mobilize and organize. Disparate militant groups coalesced under the Movement for the Emancipation of the Niger Delta (MEND) umbrella organization in 2005.
	Crisis state: December 2005–April 2006. MEND's attack on an oil pipeline signaled the first outbreak of action. This attack was followed by months of pipeline attacks, kidnappings of foreign oil workers, and a car bomb attack. MEND's public statements of its demands marked the movement's formalization.
	Institutional state: April 2006–October 2009. MEND emerged as an equal opposition player to the government after being strengthened by the crisis state. The group carried out dozens of attacks against the oil industry and publicly launched an "oil war" in the Niger Delta.
	Resolution state (co-optation): October 2009. MEND disbanded in accordance with the Amnesty Programme under a truce with the government. The peace agreement rewarded some MEND leaders with government positions but was rejected by one faction leader, who was imprisoned in 2013. The resistance continued through other militant organizations, including those that were under the MEND umbrella, and in July 2015, MEND leaders called a meeting, stirring rumors of the group's reemergence.
RUF (1991–2002)	$P > I > C \wedge I > C > N > R(f)$
	Preliminary state: 1968–1987. Rampant government corruption, state-sponsored violence, and dismal economic conditions produced restlessness and insecurity among the Sierra Leonean population, especially the unemployed youth.
	Incipient state: 1987–March 1991. The resistance began to organize when expelled students and recruits began training in Libya. The Revolutionary United Front (RUF) formed under the leadership of Foday Sankoh, and training continued in Liberia alongside the National Patriotic Front of Liberia (NPFL).

Case Study	Analysis
RUF (1991–2002)	**Crisis state:** March 1991–November 1993. RUF forces, accompanied by NPFL forces, entered Sierra Leone from Liberia, signaling the start of the insurgency. RUF operations expanded from 1992 to 1993.
	Incipient state: November 1993–1995. The new military government of Sierra Leone pushed RUF forces back into Liberia, and RUF slid back into incipiency. Internal purges and combat losses in 1993 dwindled RUF leadership and caused fragmentation. Until 1994, RUF effectively split into two operational groups.
	Crisis state: 1995–1996. RUF posed a significant threat after gaining control of the country's most critical mining sites. With RUF forces in position to attack Freetown, the vulnerable government began contracting private security and defense forces. A period of record-high violence ensued.
	Institutional state: May 1997–January 2002. On the invitation of the new government, RUF forces entered Freetown and formed the People's Army with the new junta. Demonstrating bureaucratization, RUF incorporated into the ruling junta, although internationally backed forces removed the junta in late 1997. The Lome Peace Accord was signed in July 1999 but many RUF commanders rejected it, leading to renewed violence.
	Resolution state (facilitation): January 2002. The resistance declined after the disarmament process under the 1999 peace agreement and the declared victory of government forces. RUF's lack of political objectives and popular support led its decline during peacetime, marked by peaceful elections in 2002 and 2007.
Orange Revolution (2004–2005)	P > I > C > R(s)
	Preliminary state: 1994–1999. General unrest among the population increased because of an economic downturn, unchecked government corruption, and election fraud.
	Incipient state: 1999–November 2004. Opposition to the regime coalesced after the rigged 1999 presidential election, exemplified by "Ukraine without Kuchma" protests in 2000 and 2001 in Kiev and "Our Ukraine" get-out-the-vote campaigns leading up to the 2004 elections. The first round of elections did not produce a winner, and a runoff election was planned for November 21.

131

Case Study	Analysis
Orange Revolution (2004–2005)	**Crisis state:** November 2004–December 2004. The resistance escalated as runoff elections took place. Peaceful protests broke out in Kiev shortly after Yanukovich announced the winner amid allegations of rampant election fraud. On November 22, Yushchenko declared himself president and called for continued protest. On November 27, the runoff election was declared invalid and a new election was planned.
	Resolution state (success): December 26, 2004. The third election took place, and Yushchenko won by a clear margin, signaling successful resolution of the resistance. In the absence of a unifying enemy, the resistance broke down after the election.
Solidarity (1976–1990)	P > I > C ^ I > C > R(s)
	Preliminary state: 1956–1976. This period was marked by general unrest characterized by student protests and workers strikes in response to economic downturns and government cuts. Preexisting historical and political conditions fostered a narrative of Russian oppression and singular Polish identity.
	Incipient state: 1976–1980. The resistance coalesced after the creation of the Committee for Workers' Defense and the organization of previously disparate opposition social groups (students, intellectuals, and workers). The resistance incorporated political demands into its continued economic demands. Solidarity formed in September 1980 under the leadership of Lech Walesa.
	Crisis state: 1980–1981. Action broke out in workers' strikes. The Gdansk Agreement was signed, but because the government was slow to deliver its promises, strikes and protests continued. Solidarity membership reached ten million.
	Incipient state: 1981–April 1988. The government declared martial law and outlawed Solidarity, pushing the resistance underground and back into incipiency. Membership declined, but Solidarity persisted in its underground organization. The movement maintained cohesive communications, including published newsletters, and strategy, marked by the establishment of the Provisional Coordinating Committee and other regional planning committees.

Case Study	Analysis
Solidarity (1976–1990)	**Crisis state:** April 1988–January 1989. Action escalated again in response to new hikes in food prices. By August, strikes had grown in size and expanded across the country. A televised debate between a state union leader and Walesa aired.
	Resolution state (success): January 1989. The group's normalization into a political party characterized its successful resolution. The Communist Party agreed to negotiations and allowed Solidarity to participate in upcoming elections. Solidarity swept the available seats in the election and joined the coalition government. Walesa was elected president in 1990.

[a] This description of the phases of the Iranian Coup of 1953 incorporates information from the ARIS case study as well as from Ervand Abrahamian, *A History of Modern Iran* (Cambridge: Cambridge University Press, 2008). The original case study described the military coup as spontaneous, but declassified information now shows that the CIA and MI6 were involved in planning and organizing the coup.

[b] Abrahamian, *History of Modern Iran*, 121.

[c] Ibid.

[d] If the structure allowed it, one might skip the incipient state in this case because the revolution appears to have gone directly from discontented students to a state of crisis with protests in the street.

[e] This is coded as a success, even though the new government lasted only one year. The protesters were successful in achieving their goal of overturning the government, only to fall prey to another successful revolution a year later.

Table A-2. Coded case studies without details.

Case Study	Analysis
Casebook on Insurgency and Revolutionary Warfare, volume I	
Revolution in Vietnam (1946–1954)	P > I > C > N > R(s)
Indonesian Rebellion (1945–1949)	P > I > C ^ I > C > N > R(s)
Revolution in Malaya (1948–1957)	P > I > C ^ I > C > R(e)
Guatemalan Revolution (1944)	P > I > C > N > R(s)
Venezuelan Revolution (1945)	P > I > C > N > R(l—factionalism)
Argentine Revolution (1943)	P > I > C > N > R(s)
Bolivian Revolution (1952)	P > I > C ^ I > C > N > R(e/s)
Cuban Revolution (1953–1959)	P > I > C ^ I > C > N > R(s)

Case Study	Analysis
Tunisian Revolution (1950–1954)	P > I > C > N > R(s)
Algerian Revolution (1954–1962)	P > I > C ^ I > C > N > R(s)
Revolution in French Cameroun (1956–1960)	P > I > C > N > R(p/f)
Congolese Coup (1960)	P > I > C > R(s)
Iraqi Coup (1936)	P > I > C > R(s)
Egyptian Coup (1952)	P > I > C > R(s)
Iranian Coup (1953)	P > I > C > R(s)
Iraqi Coup (1958)	P > I > C > R(s)
Sudan Coup (1958)	P > I > R(s)
Korean Revolution (1960)	P > I > C > R(s)
Chinese Communist Revolution (1927–1949)	P > I > C > N > R(s)
German Revolution (1933)	P > I > C ^ I > C > R(s)
Spanish Revolution (1936)	P > I > C > N > R(s)
Hungarian Revolution (1956)	P > I > C > R(p)
Czechoslovakian Coup (1948)	P > I > C > R(s)
Casebook on Insurgency and Revolutionary Warfare, volume II	
NPA (1969–present)	P > I > C > N ^ I(a)
FARC (1966–present)	P > I > C > N > R(i)
Shining Path (1980–1992)	P > I > C > N ^ C > R(p)
Iranian Revolution (1979)	P > I > C > R(s)
FMLN (1979–1992)	P > I > C > N > R(m)
KNLA (1949–present)	P > I > C > N ^ C > N
LTTE (1976–2009)	P > I > C > N > R(p)
PLO (1964–present)	P > I > C ^ I > C > N > R(i)
Hutu–Tutsi Genocides (1994)	P > I > C ^ I > C > R(f)
KLA (1996–1999)	P > I > C > R(s)
PIRA (1969–2001)	P > I > C > N > R(f)
Afghan Mujahidin (1979–1989)	P > I > C > N > R(s)
Viet Cong (1954–1976)	P > I > C > N ^ C ^ I > R(s)
Chechen Revolution (1991–2002)	P > I > C > N > R(r)
Hizbollah (1982–2009)	P > I > C > N
Hizbul Mujahideen (1989–present)	P > I > C > N > R(r)
EIJ (1928–2001)	P > I > C > N > R(l-encapsulation)

Case Study	Analysis
Taliban (1994–2009)	P > I > C > N ^ C > N
Al Qaeda (1988–2001)	P > I > C > N ^ C ^ I
MEND (2005–2010)	P > I > C > N > R(c)
RUF (1991–2002)	P > I > C ^ I > C > N > R(f)
Orange Revolution (2004–2005)	P > I > C > R(s)
Solidarity (1976–1990)	P > I > C ^ I > C > R(s)

APPENDIX B. ACRONYMS

AD	Democratic Action (Party)
ARIS	Assessing Revolutionary and Insurgent Strategies
CIA	US Central Intelligence Agency
EIJ	Egyptian Islamic Jihad
FARC	Revolutionary Armed Forces of Colombia
FLN	Front de Libération Nationale
FMLN	Frente Farabundo Martí para la Liberación
GOU	United Officers' Group
IRA	Irish Republican Army
KLA	Kosovo Liberation Army
KNLA	Karen National Liberation Army
KNU	Karen National Union
LTTE	Liberation Tigers of Tamil Eelam
MCP	Malayan Communist Party
MEND	Movement for the Emancipation of the Niger Delta
MTLD	Mouvement pour le Triomphe des Libertés Démocratiques
NLF	National Liberation Front
NPA	New People's Army
NPFL	National Patriotic Front of Liberia
PIRA	Provisional Irish Republican Army
PLO	Palestine Liberation Organization
RDA	Rassemblement Democratique Africain
RPF	Rwandan Patriotic Front
RUF	Revolutionary United Front
SORO	Special Operations Research Office
UN	United Nations
UPC	Union des Populations Camerounaises

BIBLIOGRAPHY

Abrahamian, Ervand. *A History of Modern Iran.* Cambridge: Cambridge University Press, 2008.

Berman, Eli, Jacob N. Shapiro, and Joseph H. Felter. "Can Hearts and Minds Be Bought? The Economics of Counterinsurgency in Iraq." *Journal of Political Economy* 119, no. 4 (2011): 766–819.

Brinton, Crane. *The Anatomy of Revolution.* Rev. ed. New York: Vintage Books, 1965.

Chenoweth, Erica, and Maria J. Stephan. "Drop Your Weapons." Foreign Affairs 93, no. 4 (July 2014): 94–106.

Christiansen, Jonathan. *Social Movements & Collective Behavior: Four Stages of Social Movements.* Research Starters Academic Topic Overviews. Ipswich, MA: EBSCO Publishing, 2009.

Cosgrove, Jonathon B., and Erin N. Hahn. *Conceptual Typology of Resistance.* Fort Bragg, NC: United States Army Special Operations Command, forthcoming.

Coy, Patrick G., and Timothy Hedeen. "A Stage Model of Social Movement Co-Optation: Community Mediation in the United States." *The Sociological Quarterly* 46, no. 3 (2005): 405–435.

Cullen, Anthony. *The Concept of Non-International Armed Conflict in International Humanitarian Law.* Cambridge: University of Cambridge, 2012.

Danelo, David J. "Exploring the Phases of Contemporary Resistance." In *Special Topics in Irregular Warfare: Understanding Resistance.* Edited by Erin Hahn. Fort Bragg, NC: United States Army Special Operations Command, forthcoming, 11–13.

Davies, James Chowning. "The J-Curve and Power Struggle Theories of Collective Violence." *American Sociological Review* 39, no. 4 (1974): 607–610.

———. "Toward a Theory of Revolution." *American Sociological Review* 27, no. 1 (1962): 5–19.

Edwards, Lyford P. *The Natural History of Revolution.* Chicago: University of Chicago Press, 1927.

Field Manual 3-24 (FM 3-24), *Counterinsurgency.* Washington, DC: Headquarters, Department of the Army, 2006.

Galulu, David and John A. Nagl. *Counterinsurgency Warfare: Theory and Practice.* Westport, CT: Creenwood Publishing Group, 2006.

Gandy, Maegen. "The Politics of Insurgency." PhD diss., University of Maryland, 2015.

Goldstone, J. "Ideology, Cultural Frameworks, and the Process of Revolution." *Theory and Society* 20, no. 4 (1991): 405–453.

Gurr, T. *Why Men Rebel.* Princeton, NJ: Princeton University Press, 1970.

Guttman, Joel M., and Rafael Reuveny. "On Revolt and Endogenous Economic Policy in Autocratic Regimes." *Public Choice* 159, no. 1 (2014) :27–52.

Hahn, Erin N., and W. Sam Lauber. *Legal Implications of the Status of Persons in Resistance.* Fort Bragg, NC: United States Army Special Operations Command, 2014.

Hopper, Rex D. "The Revolutionary Process: A Frame of Reference for the Study of Revolutionary Movements." *Social Forces* 28, no. 3 (1950): 271–272.

Higgins, Rosalyn. "International Law and Civil Conflict." In *The International Regulation of Civil Wars.* Edited by Evan Luard. New York: New York University Press, 1972.

International Committee of the Red Cross. *Commentary, Convention Relative to the Treatment of Prisoners of War.* Edited by Jean Pictet. Geneva: ICRC, 1960.

Isacson, Adam. "Colombia's Imperiled Transition." *New York Times.* April 5, 2018. https://www.nytimes.com/2018/04/05/opinion/colombia-farc-transition.html; "Colombia's FARC Officially Ceases to Be an Armed Group." *BBC News.* June 27, 2017. http://www.bbc.com/news/world-latin-america-40417207.

Jackson, Maurice, Eleanora Petersen, James Bull, Sverre Monsen, and Patricia Richmond, "The Failure of an Incipient Social Movement." *The Pacific Sociological Review* 3, no. 1 (1960): 35–40.

Jessop, Bob. "Reviewed Work: *The Natural History of Revolution* by Lyford P. Edwards." *Sociology* 6, no. 1 (1972): 130.

Juan Carlos Abella v. Argentina, Case 11.137, Report Number 55/97.

Keogh, S. "The Survival of Religious Peace Movements: When Mobilization Increases as Political Opportunity Decreases." *Social Compass* 60, no. 4 (2013): 561–578.

Knutsen, Torbjørn L., and Jennifer L. Bailey. "Review Essay: Over the Hill? The Anatomy of Revolution at Fifty." *Journal of Peace Research* 26, no. 4 (1989): 421–431.

Kotzsch, Lothar. *The Concept of War in Contemporary History and International Law.* Geneva: Libraire E. Droz, 1956.

Lauterpacht, Hersch. *Recognition in International Law.* Cambridge: Cambridge University Press, 1948.

Macionis, John J. *Sociology.* 9th ed. Upper Saddle River, NJ: Prentice Hall, 2003.

McAdam, Douglas, Sidney Tarrow, and Charles Tilly. *Dynamics of Contention.* New York: Columbia University, 2001.

Meadows, Paul. "Sequence in Revolution." *American Sociological Review* 6, no. 5 (1941): 702–709.

Miller, Frederick D. "The End of SDS and the Emergence of Weatherman: Demise through Success." In *Waves of Protest: Social Movements since the Sixties.* Edited by Jo Freeman and Victoria Johnson. Lanham, MD: Rowman & Littlefield Publishers, 1999.

Moe, Wai and Thomas Fuller. "Myanmar and 8 Ethnic Groups Sign Cease-Fire, but Doubts Remain." *New York Times.* October 15, 2015. https://www.nytimes.com/2015/10/16/world/asia/myanmar-cease-fire-armed-ethnic-groups.html.

Molnar, Andrew R. *Human Factors Considerations of Undergrounds in Insurgencies.* Washington, DC: Special Operations Research Office, 1966.

Naylor, R. T. "The insurgent Economy: Black Market Operations of Guerilla Organizations." *Crime, Law, and Social Change* 20, no. 1 (1993): 13–51.

Nyein, Nyein. "Third Session of Panglong Peace Conference Pushed Back to May." *Irrawaddy.* March 1, 2018. https://www.irrawaddy.com/news/burma/third-session-panglong-peace-conference-pushed-back-may.html.

Pettigrew, T. F. "Samuel Stouffer and relative Deprivation." *Social Psychology Quarterly* 78, no. 1 (2015): 7–24.

Powers, Robert D. "Insurgency and the Law of Nations." *JAG Journal* 16 (1962): 55–56.

Prosecutor v. Tadic. Case No. IT-94-1-AR72, Decision on Defence motion for Interlocutory Appeal on Jurisdiction. Int'l Crim. Trib. for the Former Yugoslavia, October 2, 1995.

Protocol Additional to the Geneva Conventions of 1949, and relating to the Protections of Victims of Non-International Armed Conflicts (Protocol II) art. 1(1), December 12, 1977, 1125 U.N.T.S. 609.

Sawyers, Traci M., and David S. Meyer. "Missed Opportunities: Social Movement Abeyance and Public Policy." *Social Problems* 46, no. 2 (1999): 187–206.

Solf, Waldemar A. "The Status of Combatants in Non-International Armed Conflicts under Domestic Law and Transnational Practice." *American University Law Review* 33 (1983–1984): 53–65.

Tarrow, Sidney G. *Power in Movement: Social Movements and Contentious Politics.* Rev. and updated third ed. Cambridge, NY: Cambridge University Press, 2011.

Taylor, Verta. "Social Movement Continuity: The Women's Movement in Abeyance." *American Sociological Review* 54, no. 5 (1989): 761–775.

Tse-tung, Mao. *On Guerrilla Warfare.* Urbana: University of Illinois Press, 2000.

Tun, Chit Min. "KNLA Says It Won't Attend Third Sessions of Panglong Peace Conference." *Irrawaddy.* January 9, 2018. https://www.irrawaddy.com/news/burma/knla-says-wont-attend-third-session-panglong-peace-conference.html.

Turner, Ralph H. "New Theoretical Frameworks." *The Sociological Quarterly* 5, no. 2 (1964): 122–132.

US Army Doctrine and Training Publication (ATP) 3-05, *Unconventional Warfare.* Washington, DC: Headquarters, Department of the Army, September 6, 2013.

US Central Intelligence Agency. *Guide to the Analysis of Insurgency.* Washington, DC: US Government, 2012.

Williams, R. "Relative Deprivation." In *The Idea of Social Structure: Papers in Honor of Robert K. Merton.* Edited by L. Coser. New York: Harcourt, Brace Jovanovich, 1975.

Woods, Michael, Jon Anderson, Steven Guilbert, and Suzie Watkin. " 'The Country(side) Is Angry': Emotion and Explanation in Protest Mobilization." *Social & Cultural Geography* 13, no. 6 (2012): 567–585.

Wyss, Jim. "Colombia Signs New Peace Pact with FARC Guerillas." *Miami Herald.* November 24, 2016. http://www.miamiherald.com/news/nation-world/world/americas/colombia/article116872338.html.

INDEX

A

Abella v. Argentina, 13

Abeyance: demobilization to incipience, 52–53

Action, in Meadows' phasing, 23

Adaptation, in Meadows' phasing, 23

Advanced symptoms of revolution, in Edwards' phasing, 21, 42

Afghan Mujahidin (1979-1989)

coded case study with details, 122

crisis to institutionalization to resolution in, 65–67

Algerian Revolution (1954-1962)

coded case study with details, 90–92

as failed crisis followed by repeat attempts, 72

Al Qaeda (1988-2001)

coded case study with details, 128–129

organizational destruction without outright defeat for, 69

return to crisis for, 70

The Anatomy of Revolution (Brinton), 22–23; See also Brinton, Crane

Appropriation stage, in Coy and Hedeen's phasing, 31–32

Arafat, Yasser, 73

Argentine Revolution (1943)

coded case study with details, 86–87

crisis to institutionalization to resolution in, 65, 68

ARIS Casebook on Insurgency and Revolutionary Warfare, 36, 59

Assessing Revolutionary and Insurgent Strategies (ARIS), 1

Casebook on Insurgency and Revolutionary Warfare, 36, 59

Conceptual Typology of Resistance, 1

Legal Implications of the Status of Persons in Resistance, 12–16

phasing construct, 1–2, 10

previous study on contemporary phases of resistance, 2–9

Special Topics in Irregular Warfare: Understanding Insurgency, 9–10

Assimilation or transformation stage, in Coy and Hedeen's phasing, 32

ATP 3-05; See US Army Doctrine and Training Publication (ATP) 3-05

B

Balked disposition, 37

Belligerency, 13, 15–16

Berman, Eli, 19

Bolivian Revolution (1952)

coded case study with details, 87–88

as failed crisis followed by repeat attempts, 72

Bourgeois-nationalist shortcut pattern, 4, 5

Brinton, Crane, 22–23, 37, 42, 46

Bureaucratization, 49–52

C

Case studies; See also *individual case studies*

coded and grouped according to path through, 60–62

coding key for, 60

coding progression through states, 59–60

of crisis to institutionalization to resolution, 65–68

of failed crises followed by repeat attempts, 72–74

of organizational destruction without outright defeat, 69–71

as proof of states of resistance concept, 59–74

of resolution without crises, 62–63

of short crises with decisive resolutions, 63–65

Chechen Revolution (1991-2002)

    coded case study with details, 124

    crisis state in, 47–48

    crisis to institutionalization to resolution in, 66, 67

Chinese Communist Revolution (1927-1949)

    coded case study with details, 104–105

    crisis to institutionalization to resolution in, 65, 67

CIA; See US Central Intelligence Agency

CIA Guide; See *Guide to the Analysis of Insurgency* (CIA)

Clash of radical revolutionary factions, in Brinton's phasing, 22–23

Coalescence, 41–45

Coding, case studies

    grouped according to path, 60–62

    key for, 60

    progression through states, 59–60

Collective behavior and conflict constructs, 32–33

Conceptual Typology of Resistance (ARIS), 1

Congolese Coup (1960)

    coded case study with details, 94–96

    short crisis with decisive resolution in, 63, 64

Contentious mobilization, mechanisms for, 28

Contentious politics, mechanisms of, 28

Co-optation

    code for, 81

    in Coy and Hedeen's phasing, 32, 58

    in Miller's phasing, 27

    as resolution state, 57–58

Counterinsurgency Warfare (Galula), 2, 4–6

"The Country(side) Is Angry" (Woods et al.), 33–35

Coy, Patrick G., 31–32, 58

Crisis stage

    and Brinton's phasing, 46

    in Edwards' phasing, 21, 46

    in Meadows' phasing, 23, 46

Crisis state

    code for, 81

    as formalization and outbreak of action, 45–49

Cuban Revolution (1953-1959)

    coded case study with details, 88–89

    as failed crisis followed by repeat attempts, 72

Czechoslovakian Coup (1948)

    coded case study with details, 108–109

    short crisis with decisive resolution in, 63, 64

D

Davies, James Chowning, 38

Decline

    as resolution state, 53

    in social movement theory, 53

    UCLA authors on, 30

Demobilization
   in abeyance state, 52–53
   mechanisms for, 28
Dormancy, 52

E

Economics literature review, 16–21
   Berman, Shapiro, and Felter's mathematical model and game theory, 19
   Guttman and Reuveny's game theory model, 19
   Naylor's analysis, 16–21
Edwards, Lyford P., 21–22, 37, 42, 46, 59
Egyptian Coup (1952)
   coded case study with details, 97–99
   short crisis with decisive resolution in, 63, 64
Egyptian Islamic Jihad (EIJ) (1928-2001)
   coded case study with details, 126–127
   crisis to institutionalization to resolution in, 66, 67
Emergence state, 37
Emergent norm versus contagion theory construct, 33
Emotions of activists, evolution of, 33–34
Encapsulation
   code for, 81
   as failure state, 57
Establishment with mainstream
   code for, 81
   as resolution state, 58
Exhaustion
   code for, 81
   as resolution state, 59

External funding, 18–19

F

Facilitation
   code for, 81
   as resolution state, 55
Factionalism
   code for, 81
   as failure state, 57
Failure
   code for, 81
   in Miller's phasing, 28, 57
   as resolution state, 57
   UCLA authors on, 31
FARC; See Revolutionary Armed Forces of Colombia (1966-present)
Felter, Joseph H., 19
Financing, stages of, 17–18, 20
FMLN; See Frente Farabundo Martí para la Liberación Nacional (1979-1992)
Formalization, as crisis state, 45–49
Formal stage, in Hopper's phasing, 25–26, 46–47
Freeman, Jo, 26
French Cameroun, revolution in (1956-1960)
   coded case study with details, 92–94
   crisis to institutionalization to resolution in, 66, 67
Frente Farabundo Martí para la Liberación Nacional (FMLN) (1979-1992)
   coded case study with details, 113–114
   crisis to institutionalization to resolution in, 66, 67
   establishment with the mainstream, 58

G

Galula, David, 2, 4–6, 39

German Revolution (1933)

    coded case study with details, 105–106

    as failed crisis followed by repeat attempts, 72

Griffith, Samuel, 3

Guatemalan Revolution (1944)

    coded case study with details, 84–85

    crisis to institutionalization to resolution in, 65, 67–68

Guerrilla Warfare (Mao), 43

Guide to the Analysis of Insurgency (CIA), 28–30

    crisis stage, 47

    incipience, 43

    institutional state, 49

    preliminary phase, 39

    on resolution, 53

Gurr, T., 38

Guttman, Joel M., 19

Guzman, Abimael, 69

H

Hedeen, Timothy, 31–32, 58

Highly publicized failures

    code for, 81

    results of, 57

Hitler, Adolf, 72

Hizbollah (1982-2009)

    coded case study with details, 125

    crisis to institutionalization to resolution in, 66, 67

    institutional state of, 50

Hizbul Mujahideen (1989-present)

    coded case study with details, 126

    crisis to institutionalization to resolution in, 66, 67

Hopper, Rex D., 23–26, 37, 42, 46–47, 49, 54

Human Factors Considerations of Undergrounds in Insurgencies (SORO), 2, 6–7

Hungarian Revolution (1956), 64

    coded case study with details, 107–108

    short crisis with decisive resolution in, 64–65

Hutu–Tutsi genocides (1994)

    coded case study with details, 118–119

    as failed crisis followed by repeat attempts, 72, 73

I

Imminent versus interactive determination construct, 33

Inception and engagement stage, in Coy and Hedeen's phasing, 31–32

Incipient stage

    coalescence in, 41–45

Incipient state

    as coalescence, 41–45

    code for, 81

    regressing from, 62–63

Incorporating other leadership, failure in, 57

    code for, 81

Incubation stage; See also Preliminary stage; Preliminary state

    in ARIS framework, 37–41

    in Meadows' phasing, 23

Indonesian Rebellion (1945-1949)

    coded case study with details, 82–83

as failed crisis followed by repeat attempts, 72

Institutionalization

code for, 81

as resolution state, 54

Institutional stage

in Hopper's phasing, 26, 54

as legalization and societal organization, 26

Institutional state

as bureaucratization, 49–52

code for, 81

Insurgency, 13, 15

International humanitarian law, 13–16

Iranian Coup (1953)

coded case study with details, 99–101

short crisis with decisive resolution in, 63, 64

Iranian Revolution (1979)

coded case study with details, 113

incipient state in, 43–44

short crisis with decisive resolution in, 63–65

Iraqi Coup (1936)

coded case study with details, 96–97

short crisis with decisive resolution in, 63, 64

Iraqi Coup (1958)

coded case study with details, 101–102

short crisis with decisive resolution in, 63, 64

J

Jackson, Maurice, 57; See also University of California, Los Angeles study

Johnson, Victoria, 26

K

Karen National Liberation Army (KNLA) (1949-present)

coded case study with details, 114–115

institutional state of, 50–51

organizational destruction without outright defeat for, 69, 70

reinstitutionalization of, 70, 71

KLA; See Kosovo Liberation Army (1996-1999)

KNLA; See Karen National Liberation Army (1949-present)

Korean Revolution (1960)

coded case study with details, 103–104

short crisis with decisive resolution in, 63–65

Kosovo Liberation Army (KLA) (1996-1999)

coded case study with details, 120–121

short crisis with decisive resolution in, 63, 65

L

Ladder of emotions, 34

Latent state, 37

Law literature review, 12–16

Legal Implications of the Status of Persons in Resistance (ARIS), 12–16

Liberation Tigers of Tamil Eelam (LTTE) (1976-2009)

coded case study with details, 116–117

crisis to institutionalization to resolution in, 66, 67

preliminary state of, 41

Literature review, 11–35
early analysis, 12
economics literature, 16–21
law literature, 12–16
political science and social
movement theory literature,
21–35
process for, 11
results of, 11–12
LTTE; See Liberation Tigers of Tamil
Eelam (1976-2009)

M

Macionis, John J., 58
Malaya, revolution in (1948-1957)
coded case study with details, 83–84
as failed crisis followed by repeat
attempts, 72, 73
Mao, 2–4, 7
crisis state, 47
incipient state, 43
institutional state, 49
preliminary state, 39
resolution states, 53
Meadows, Paul, 23, 37, 42, 46, 49
Mechanisms
for contentious mobilization, 28
of contentious politics, 28
for demobilization, 28
Mechanisms of contentious politics, 28
MEND; See Movement for the
Emancipation of the Niger Delta
(2005-2010)
Methodology of study, 9–10
Meyer, David S., 52–53
Miller, Frederick D., 26–28, 55–58
"Missed Opportunities" (Sawyers and
Meyer), 52–53

Mobilization, contentious, mechanisms
for, 28
Movement for the Emancipation of
the Niger Delta (MEND) (2005-
2010)
coded case study with details,
129–130
crisis to institutionalization to
resolution in, 66, 67
Muslim Brotherhood in Egypt, 50

N

The Natural History of Revolution
(Edwards), 21–22; See also
Edwards, Lyford P.
Naylor, R. T., 16–21
New People's Army (NPA)
(1969-present)
coded case study with details,
109–110
crisis to institutionalization to
resolution in, 66, 67
"New Theoretical Frameworks"
(Turner), 32–33
Noninternational armed conflict, 15
Nonviolent resistance, 16
lawful measures, 13–14
unlawful measures, 13, 14
Normality, in Edwards' phasing, 21, 22
NPA; See New People's Army
(1969-present)

O

Objective of study, 9–10
Open insurgency stage, in CIA's
phasing, 29
Orange Revolution (2004-2005)
coded case study with details,
131–132
incipient state in, 45

short crisis with decisive resolution
in, 63

success of, 56

Orthodox communist pattern, in
Galula phasing, 4–5

Outbreak of action, as crisis state,
45–49

Outbreak of revolution, in Edwards'
phasing, 21

P

Palestine Liberation Organization
(PLO) (1964-present), 73

coded case study with details,
117–118

as failed crisis followed by repeat
attempts, 72

resolution by institutionalization
for, 54

Parasitical fund-raising, 17, 20

Pettigrew, Thomas F., 38–39

Phasing constructs

ARIS, 1–2 (See also States of
resistance framework)

ATP 3-05, 2, 8

FM 3-24, 8

of Galula, 4–6

of Mao, 2, 3, 7

review of (See Literature review)

of SORO, 6–7

US Army Field Manual 3-24, 2–3

PIRA; See Provisional Irish Republican
Army (1969-2001)

PLO; See Palestine Liberation
Organization (1964-present)

Political science and social movement
theory literature review, 21–35

Brinton, *The Anatomy of Revolution*,
22–23, 37, 42, 46

CIA, *Guide to the Analysis of
Insurgency*, 28, 39, 43, 47, 49, 53

Coy and Hedeen, "A Stage Model of
Social Movement Co-Optation,"
31–32, 58

Edwards, *The Natural History of
Revolution*, 21–22, 37, 42, 46, 59

Hopper, "The Revolutionary
Process," 23–26, 42, 46–47, 49,
54

Meadows, "Sequence in
Revolution," 23, 37, 42, 46, 49

Miller, in Waves of Protest, 26–28,
55–58

Tarrow, *Power in Movement*, 28,
53–55, 59

Turner, "New Theoretical
Frameworks," 32–33

University of California, Los
Angeles study, 30–31

Woods et al., "The Country(side) Is
Angry," 33–35

Popular stage, in Hopper's phasing,
24–25

Postcrisis stage, in Meadows' phasing,
23, 49

Power in Movement (Tarrow), 28; See
also Tarrow, Sidney G.

Precrisis stage, in Meadows' phasing,
23

Predatory funding, 17, 20

Preexisting network, failure in, 57
code for, 81

Preinsurgency stage, in CIA's phasing,
29, 39

Preliminary stage

in Brinton's phasing, 22, 37

in Edwards' phasing, 21, 37

in Hopper's phasing, 24

in Meadows' phasing, 23, 37

Preliminary state
    code for, 81
    as incubation, 37–41
    regressing from incipiency to, 62–63
    and resolution without crises, 62
Process resolution versus unfolding construct, 32–33
Program for support, lack of, 57
    code for, 81
Progression in phasing, code for, 81
Provisional Irish Republican Army (PIRA) (1969-2001)
    coded case study with details, 121
    crisis state in, 48
    crisis to institutionalization to resolution in, 66, 67
    resolution through facilitation for, 55–56
Public funding of resistance, 19

R

Radicalization
    code for, 81
    as resolution state, 53–54
Rapport, 37
Rebellion, 13–15
Recovery stage, in Brinton's phasing, 23
Regulation and response stage, in Coy and Hedeen's phasing, 32
Relative deprivation, 37–39
Repression
    code for, 81
    in Miller's phasing, 27
    as resolution state, 55
Resistance, 1

Resistance movements
    phases of, 1 (See also States of resistance framework)
    social movements and, 45–46
    variables in, 9
Resolution stage, in CIA phasing, 29
Resolution state
    decline, 53
    facilitation, 55
    radicalization, 53–54
    without crises, 62–63
Resolution states, 53–59
    code for, 81
    establishment with mainstream, 58
    exhaustion, 59
    failure, 57
    institutionalization, 53
    repression, 55
    success, 56
Reuveny, Rafael, 19
Reversal in phasing, code for, 81
Revolutionary Armed Forces of Colombia (FARC) (1966-present)
    coded case study with details, 110–111
    crisis to institutionalization to resolution in, 66, 67
    institutional state of, 51–52
"The Revolutionary Process" (Hopper), 23–26; See also Hopper, Rex D.
Revolutionary United Front (RUF) (1991-2002)
    coded case study with details, 130–131
    as failed crisis followed by repeat attempts, 72S

S

Sawyers, Traci M., 52–53

"Sequence in Revolution" (Meadows), 23; See also Meadows, Paul

Shapiro, Jacob N., 19

Shining Path (1980-1992)

    coded case study with details, 111–112

    incipient state in, 44–45

    organizational destruction without outright defeat for, 69–70

    return to crisis for, 69

Social movements, resistance movements and, 45–46

Social movement theory

    decline in, 53

    literature reviews (See Political science and social movement theory literature review)

Solidarity (1976-1990)

    coded case study with details, 132–133

    as failed crisis followed by repeat attempts, 72, 73

    preliminary state of, 39–40

SORO; See Special Operations Research Office

South African antiapartheid resistance, 14

Spanish Revolution (1936)

    coded case study with details, 106–107

    crisis to institutionalization to resolution in, 66, 67

Special Operations Research Office (SORO), 1, 43

    and crisis state, 47

    Human Factors Considerations of Undergrounds in Insurgencies, 2, 6–7

and preliminary state, 39

Special Topics in Irregular Warfare: Understanding Insurgency (ARIS), 9–10

"A Stage Model of Social Movement Co-Optation" (Coy and Hedeen), 31–32

States of resistance framework, 35–59

    abeyance: demobilization to incipience, 52–53

    case studies as proof of concept for, 59–74 (See also Case studies)

    crisis state: formalization and outbreak of action, 45–49

    incipient state: coalescence, 41–45

    institutional state: bureaucratization, 49–52

    preliminary state: incubation, 37–41

    resolution states, 53–59

Success

    code for, 81

    in Miller's phasing, 27, 56

    as resolution state, 56

Sudan Coup (1958)

    coded case study with details, 102–103

    resolution without crisis in, 63

Symbiotic fund-raising, 18, 20

T

Taliban (1994-2009)

    coded case study with details, 127–128

    organizational destruction without outright defeat for, 69

    preliminary state of, 40–41

    reinstitutionalization of, 70

Tarrow, Sidney G., 28, 53–55, 59

Taylor, Verta, 53

Tunisian Revolution (1950-1954)

    coded case study with details, 89–90

    crisis to institutionalization to resolution in, 65, 68

Turner, Ralph, 32–33

U

Ukrainian Revolution, short crisis with decisive resolution in, 64–65

University of California, Los Angeles study, 30–31; See also Jackson, Maurice

US Army Doctrine and Training Publication (ATP) 3-05, 2–4, 8

    and crisis stage, 47

    and preliminary state, 39

US Army Field Manual 3-24, 2–3, 8

US Central Intelligence Agency (CIA), 28; See also *Guide to the Analysis of Insurgency* (CIA)

US civil rights movement, 14

V

Venezuelan Revolution (1945)

    coded case study with details, 85–86

    crisis to institutionalization to resolution in, 66, 68

Viet Cong (1954-1976)

    coded case study with details, 123

    crisis state in, 48–49

    organizational destruction without outright defeat for, 69

    return to crisis for, 70

Vietnam, revolution in (1946-1954)

    coded case study with details, 82

    crisis to institutionalization to resolution in, 65

W

Waves of Protest (Freeman and Johnson), 26

Williams, R., 39

Woods, Michael, 33–35

Z

Zones of contention stage, 17

Zones of control stage, 18

Zones of exclusion stage, 17

www.ingramcontent.com/pod-product-compliance
Lightning Source LLC
Chambersburg PA
CBHW052112020426

42335CB00021B/2733